# TABLE OF CO

## Teacher Tips

- Review instructions for students to ensure understanding of the questions.
- Encourage students to complete the answers they know how to do first.
- Create a data management word wall to reinforce math vocabulary.

## Black Line Masters

Create your own worksheets using the black line masters available in this book to review, clarify or reinforce data management concepts.

## Rubric And Checklists

Use the rubric and grade checklists to help monitor and assess students' learning.

# SORTING FUN

Cut out and paste each item into the correct box.

## VEGETABLES

apple

lettuce

cauliflower

banana

cucumbers

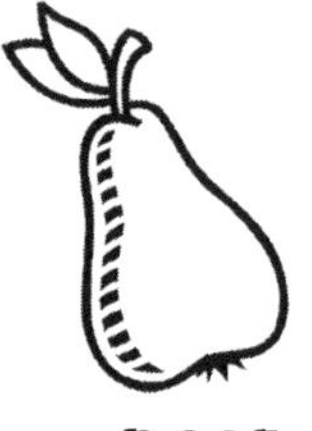

pear

strawberries

carrot

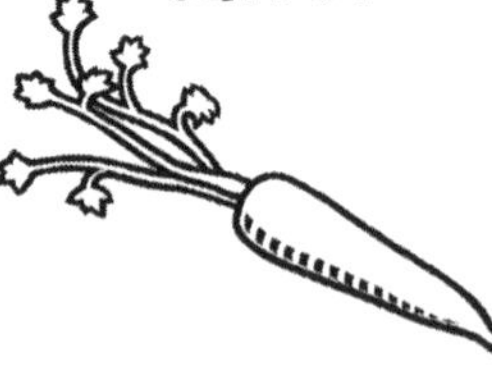

# SORTING FUN

Cut out and paste each item into the correct box.

| BIRDS | FISH |
| --- | --- |
| | |

# SORTING FUN

Cut out and paste each item into the correct box.

# READING TALLY CHARTS

Use the tally chart to answer the questions.

## FAVOURITE FISH

Count. How many?

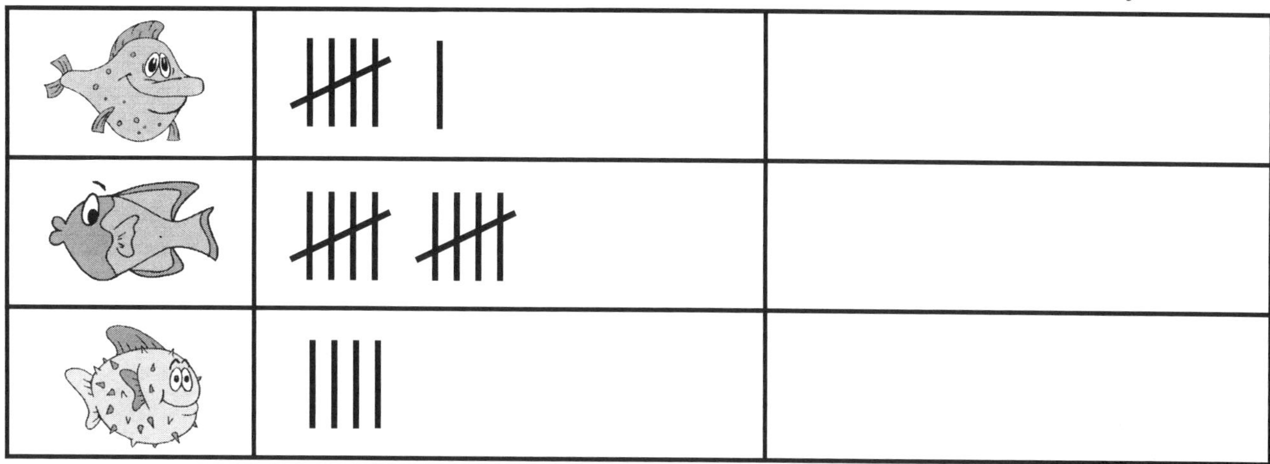

1. Circle the <u>most</u> popular fish.   

2. Circle the <u>least</u> popular fish.   

**BRAIN STRETCH:** Circle the correct answer.

How are these things alike?

A. You can eat them. B. You can wear them. C. You can write with them.

# READING TALLY CHARTS

Mr. Cummin's class made a tally chart to show the results of their favourite activity survey. Use the tally chart to answer the questions.

## FAVOURITE ACTIVITY

Count. How many?

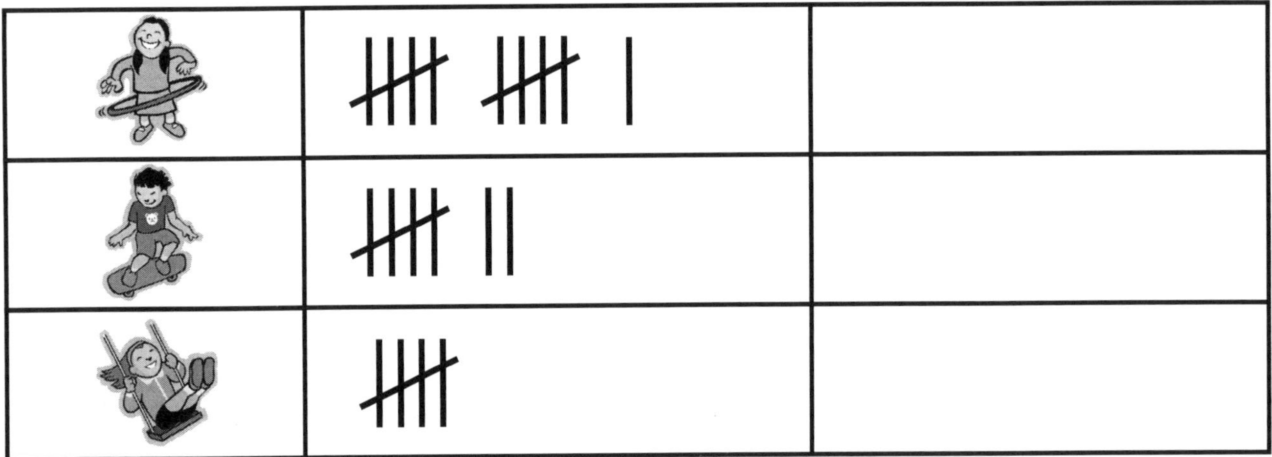

1. Circle the <u>most</u> popular activity.

2. Circle the <u>least</u> popular activity.

**BRAIN STRETCH:** Which item doesn't belong with this set?

A. 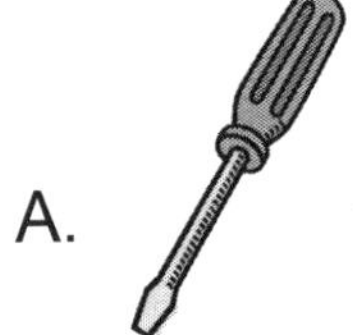

B.

C. 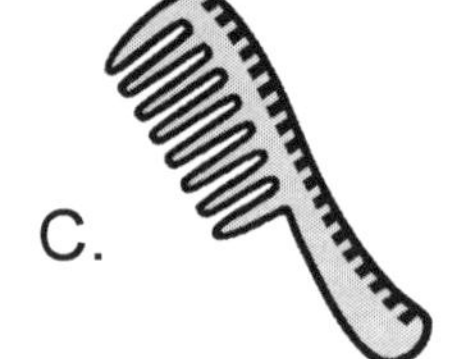

D. 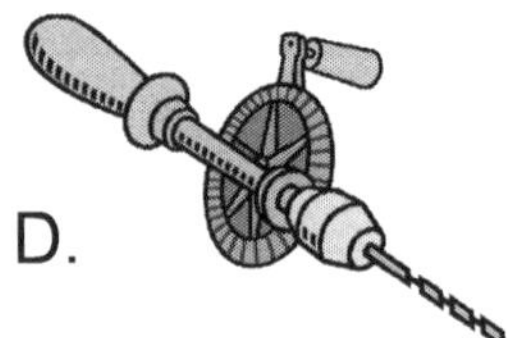

# READING TALLY CHARTS

Use the tally chart to answer the questions.

## FAVOURITE VEGETABLE

Count. How many?

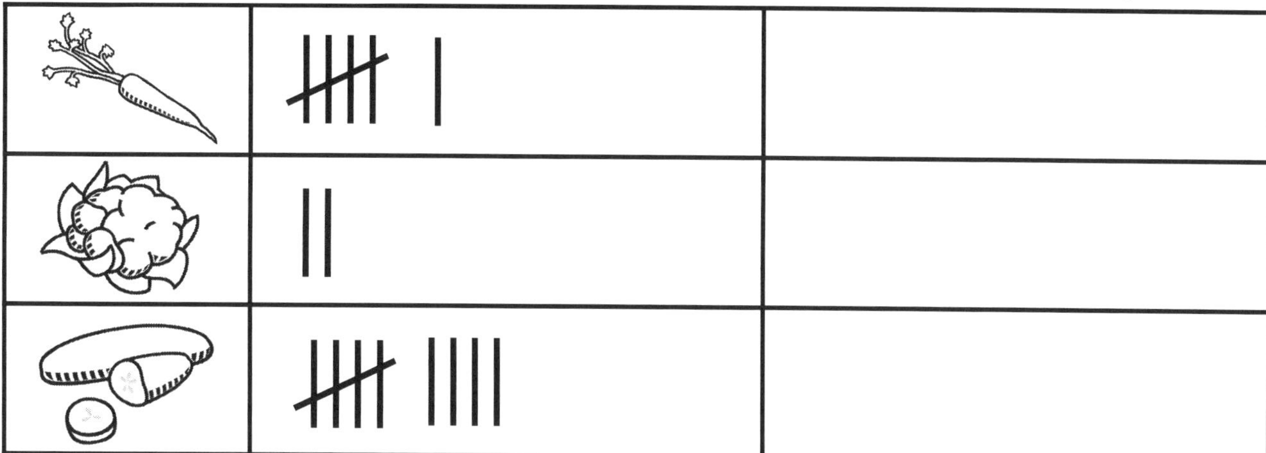

1. Circle the <u>most</u> popular vegetable.

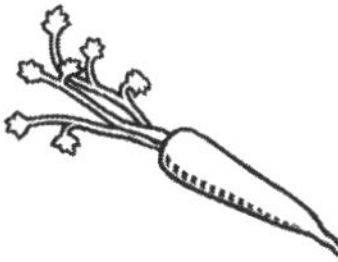  

2. Circle the <u>least</u> popular vegetable.

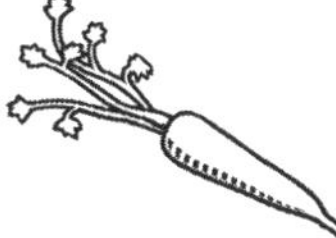  

**BRAIN STRETCH:** Which item goes with this set?

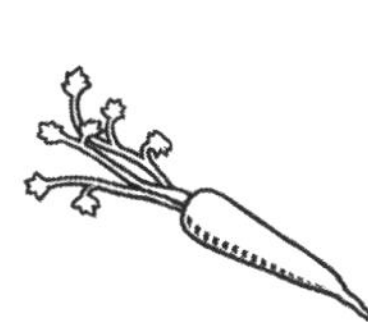  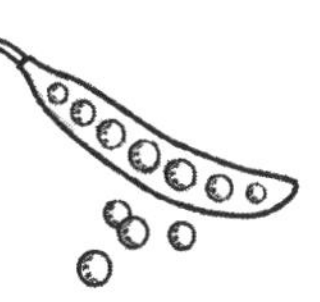

A. 

B. 

C. 

# READING TALLY CHARTS

Here are the results of a favourite colour survey. Complete the chart.

| Colour | Tally | Number |
|---|---|---|
| Red | 卌 \|\| | |
| Blue | 卌 卌 \|\|\| | |
| Green | \|\|\|\| | |
| Purple | 卌 \|\|\| | |

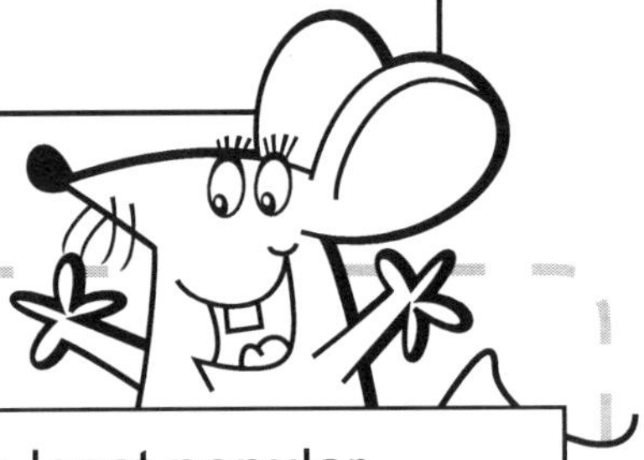

**BRAIN STRETCH:** Answer the questions.

| | | | |
|---|---|---|---|
| **1.** | What was the most popular colour? | **2.** | What was the least popular colour? |
| **3.** | What colour had 8 votes? | **4.** | How many votes were there for red and blue altogether? |
| **5.** | How many people voted altogether? | **6.** | List the colours from the most votes to the least votes. |

# EXPLORING TALLY CHARTS

Here are the results of a favourite breakfast food survey. Complete the chart.

| Favourite Breakfast Foods | Tally | Number |
|---|---|---|
| Cereal | | 3 |
| Eggs | | 15 |
| Pancakes | | 21 |
| Grilled Cheese | | 15 |

**BRAIN STRETCH:** Answer the questions.

| | | | |
|---|---|---|---|
| **1.** | What was the most popular breakfast food? | **2.** | How many people like cereal and pancakes altogether? |
| **3.** | Which breakfast food did people like the same? | **4.** | How many fewer peope liked cereal than eggs? |
| **5.** | Which breakfast food had 3 votes? | **6.** | How many people liked grilled cheese and cereal altogether? |

# EXPLORING BAR GRAPHS

Here is a data table to show what summer activities students in Ms. Richardson's class prefer. Complete the chart and bar graph to show the information. Make sure to label!

| Summer Activity | Number of Votes | Number |
|---|---|---|
| Swimming | 𝍸 𝍸 | |
| Baseball | 𝍸 \|\|\|\| | |
| Soccer | 𝍸 𝍸 \| | |
| Going to the Beach | 𝍸 \|\| | |

______________________________

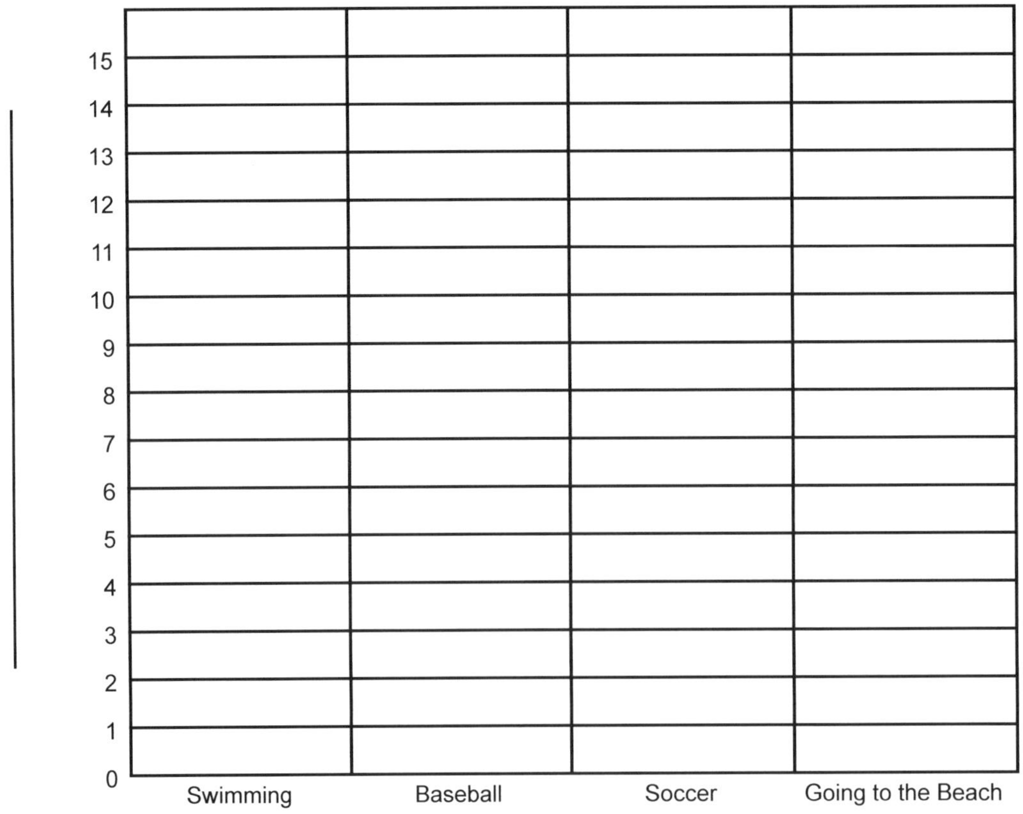

15
14
13
12
11
10
9
8
7
6
5
4
3
2
1
0

Swimming Baseball Soccer Going to the Beach

______________________________

# EXPLORING BAR GRAPHS

Here is a data table to show the number of different coins David has in his piggy bank. Complete the chart and bar graph to show the information. Make sure to label!

| Type of Coins | Number of Coins | Number |
|---|---|---|
| pennies | 卌 卌 卌 | |
| nickels | 卌 \|\|\|\| | |
| dimes | 卌 卌 卌 \| | |
| quarters | 卌 \| | |
| loonies | 卌 | |

______________________________

16
15
14
13
12
11
10
9
8
7
6
5
4
3
2
1
0

pennies nickels dimes quarters loonies

______________________________

# MAKE A TALLY CHART

Use the information from the favourite pet survey to complete the tally chart.

## FAVOURITE PET SURVEY

| Person | Favourite Pet |
|---|---|
| Neeha | fish |
| Josh | fish |
| Somie | cat |
| John | dog |
| Sarah | dog |
| Doug | cat |
| Cindy | fish |
| Patricia | bird |
| Ross | dog |

| Pet | Number of Votes |
|---|---|
| fish | |
| bird | |
| cat | |
| dog | |

**BRAIN STRETCH:** Answer the questions.

| | | | |
|---|---|---|---|
| **1.** | Which two kinds of pets have the same number of votes? | **2.** | Which kind of pet has the least number of votes? |

# MAKE A TALLY CHART

Use the information from the favourite drink survey to complete the tally chart.

## FAVOURITE DRINK SURVEY

| Person | Favourite Drink |
|---|---|
| Mary | juice |
| Paul | lemonade |
| Katie | soda |
| John | milk |
| Ivy | juice |
| Jeff | lemonade |
| Linda | milk |
| Patricia | milk |
| Steven | juice |
| Elizabeth | milk |

| Drink | Number of Votes |
|---|---|
| juice | |
| lemonade | |
| soda | |
| milk | |

**BRAIN STRETCH:** Answer the questions.

| | | | |
|---|---|---|---|
| **1.** | Which drink has 3 votes? | **2.** | Which two drinks have the same number of votes? |

# READING CHARTS

Ben went fishing. Look at the chart to see the number of fish Ben caught from a Monday to a Friday.

| Day of the Week | Monday | Tuesday | Wednesday | Thursday | Friday |
|---|---|---|---|---|---|
| Number of Fish Caught | 3 | 6 | 9 | 12 | 15 |

**BRAIN STRETCH:** Answer the questions.

| | | | |
|---|---|---|---|
| **1.** | On what day did Ben catch the most number of fish? | **2.** | How many fish did Ben catch on Thursday? |
| **3.** | On what day did Ben catch the least number of fish? | **4.** | What is the difference between the most number of fish Ben caught and the least number of fish? |
| **5.** | How many fish did Ben catch on Tuesday and Wednesday? | **6.** | How many fish do you think Ben will catch on Saturday? |

# READING CHARTS

Here is a chart to show the number of students and desks in each grade three class.

| Room Data | Ms. Apor | Mr. Patel | Mr. Poulos | Ms. Chin |
|---|---|---|---|---|
| Students | 22 | 25 | 24 | 25 |
| Desks | 25 | 25 | 21 | 27 |

**BRAIN STRETCH:** Answer the questions.

| | | | |
|---|---|---|---|
| **1.** | Who needs more desks in their room? | **2.** | How many students does Ms. Apor have? |
| **3.** | Which two teachers have the same number of students? | **4.** | Which teacher has the least number of students? |
| **5.** | Which teacher has more desks than they need? | **6.** | Which teacher needs 3 more desks? |

# READING PICTOGRAPHS

Use the pictograph to answer the questions.

1. How many?  ______  ______  ______

2. Circle the <u>most</u> popular toy.   

3. Circle the <u>least</u> popular toy.   

# READING PICTOGRAPHS

Ms. Apor's class made a pictograph to show the results of their favourite footwear survey. Use the pictograph to answer the questions.

## FAVOURITE FOOTWEAR

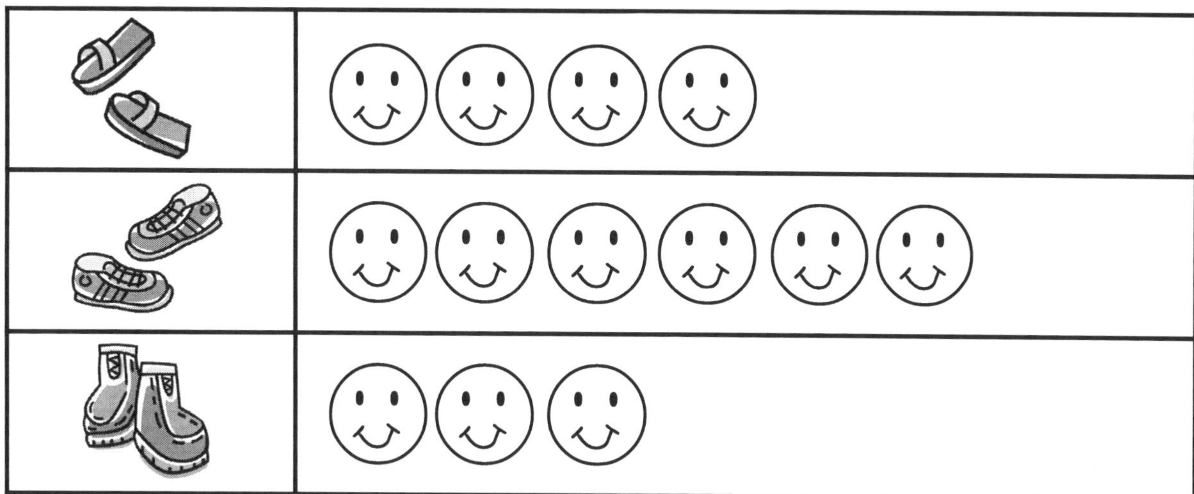

1. How many?  ______  ______  ______

2. Circle the most popular footwear.   

3. Circle the least popular footwear.   

**BRAIN STRETCH:** Count the tally marks.

How many?

| | |
|---|---|
|  |  |

# READING PICTOGRAPHS

Use the pictograph to answer the questions.

## FAVOURITE PET

1. How many?   

2. Circle the most popular pet.   

3. Circle the least popular pet.   

**BRAIN STRETCH:** Count the tally marks.

How many?

# READING PICTOGRAPHS

Use the pictograph to answer the questions.

## FAVOURITE CAKE

1. How many? 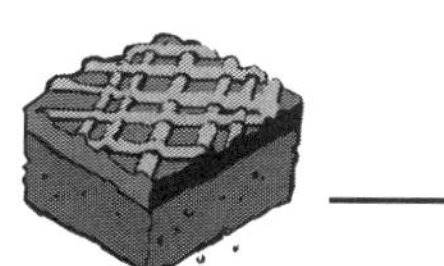 ______  ______  ______

2. Circle the <u>most</u> popular cake.   

3. Circle the <u>least</u> popular cake. 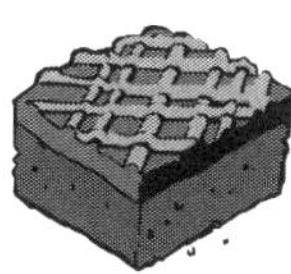  

4. How many people voted in the survey altogether? ____________

# READING PICTOGRAPHS

Use the pictograph to answer the questions.

## BRAIN STRETCH: Answer the questions.

1. How many?  ______  ______  ______

2. Circle the <u>most</u> popular indoor activity.   

3. Circle the <u>least</u> popular indoor activity.   

4. How many people voted in the survey altogether? ______________

# READING TALLY CHARTS

Complete the tally chart.

| | Number | Tally |
|---|---|---|
| | **8** | |
| | **16** | |
| | **20** | |

**BRAIN STRETCH:** Answer the questions.

1. Circle the <u>most</u> popular transportation.

2. Circle the <u>least</u> popular transportation.

3. How many people liked  <u>more than</u>  ______ ?

4. How many people liked  <u>less than</u>  ______ ?

# READING PICTOGRAPHS

Here is a pictograph that shows the type of transportation students use to go to school.

## TYPE OF TRANSPORTATION

| | |
|---|---|
| School Bus | 🯅🯅🯅🯅 |
| Walk | 🯅🯅🯅🯅🯅🯅🯅🯅 |
| Car | 🯅🯅🯅🯅🯅 |
| Bicycle | 🯅🯅 |

🯅 = 2 students

**BRAIN STRETCH:** Answer the questions.

| | | | |
|---|---|---|---|
| **1.** | How many students ride the school bus? | **2.** | How many more students get a car ride than ride a bicycle? |
| **3.** | How many students walk? | **4.** | How many students ride their bike? |
| **5.** | How many fewer students ride the school bus than walk? | **6.** | How many choices of transportation are in the pictograph? |

# READING PICTOGRAPHS

Here is a pictograph to show the results of a favourite drink survey.

## FAVOURITE DRINKS

| Drink | |
|---|---|
| Lemonade | 🥤🥤🥤🥤🥤🥤🥤 |
| Milk | 🥤🥤🥤 |
| Orange Juice | 🥤🥤🥤🥤🥤 |

🥤 = 10 people

**BRAIN STRETCH:** Answer the questions.

1. What drink do people like the most? ___________
2. What drink do people like the least? ___________
3. How many people chose milk? ___________
4. How many people chose orange juice? ___________
5. How many people chose lemonade? ___________
6. How many people voted in this survey? ___________
7. How many more people chose lemonade than milk? ___________
8. How many people chose orange juice and milk? ___________

# READING PICTOGRAPHS

Megan took a survey on favourite ice cream flavours of her classmates. Here are the results of the survey.

## FAVOURITE ICE CREAM FLAVOURS

| | |
|---|---|
| Strawberry | 🍦🍦🍦🍦🍦🍦🍦🍦🍦🍦🍦 |
| Chocolate Chip | 🍦🍦🍦🍦🍦🍦 |
| Vanilla | 🍦🍦🍦🍦🍦🍦🍦🍦🍦🍦 |
| Chocolate | 🍦🍦🍦🍦🍦 |

🍦 = 4 people

**BRAIN STRETCH:** Answer the questions.

1. What flavour is the most popular? ______________________
2. What flavour is the least popular? ______________________
3. How many students chose vanilla? ______________________
4. How many fewer students chose chocolate over vanilla? ____________
5. How many students chose either strawberry or vanilla? ____________
6. How many people chose chocolate? ______________________
7. How many people chose chocolate chip? ______________________
8. How many more poeple chose strawberry over chocolate chip? ________
9. How many students chose strawberry? ______________________
10. How many people voted altogether? ______________________

# READING PICTOGRAPHS

Here is a pictograph to show the results of a favourite cookie survey.

## FAVOURITE COOKIE FLAVOURS

| Chocolate Chip | |
|---|---|
| Oatmeal Raisin | |
| Gingerbread | |
| Sugar | |

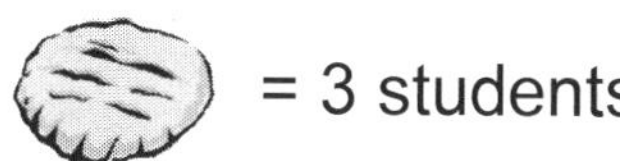 = 3 students

**BRAIN STRETCH:** Answer the questions.

| | | | |
|---|---|---|---|
| **1.** | How many students chose chocolate chip? | **2.** | How many students chose oatmeal raisin? |
| **3.** | How many students chose gingerbread? | **4.** | Which cookie is the most popular? |
| **5.** | How many more students chose oatmeal raisin than gingerbread? | **6.** | List the cookies from the most favourite to the least favourite. |

# READING PICTOGRAPHS

Here is a pictograph to show how many cookies a group of children sold during a school bake sale.

## NUMBER OF COOKIES SOLD

| | |
|---|---|
| Tom | |
| Lisa | |
| Jared | |
| Miguel | |
| Sarah | |

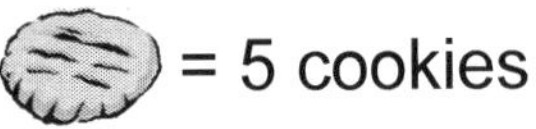 = 5 cookies

**BRAIN STRETCH:** Answer the questions.

1. Who sold the most cookies?
2. How many more cookies did Miguel sell than Jared?
3. How many cookies did Tom sell?
4. How many more cookies did Tom sell than Sarah?
5. How many cookies did Sarah sell?
6. How many cookies were sold altogether?

# EXPLORING BAR GRAPHS

Count the pictures and complete the bar graph.

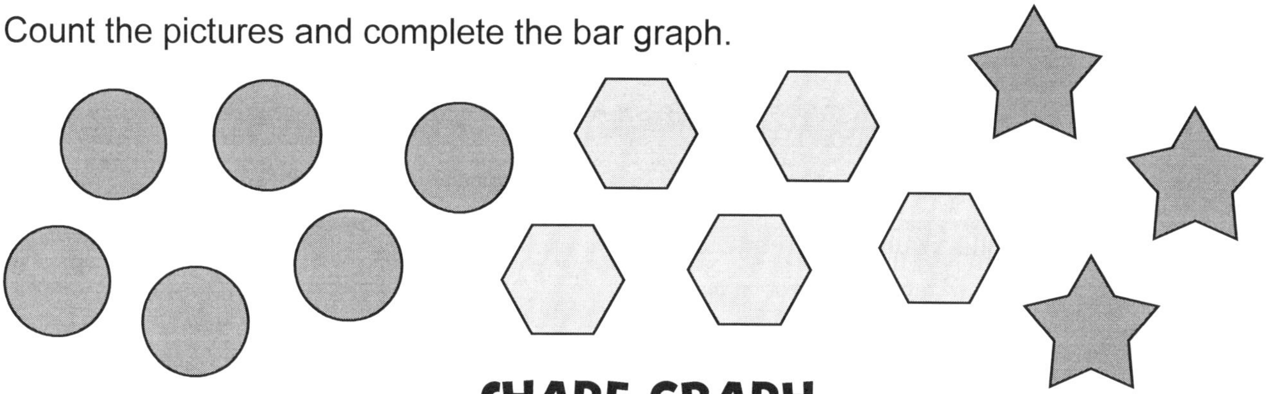

## SHAPE GRAPH

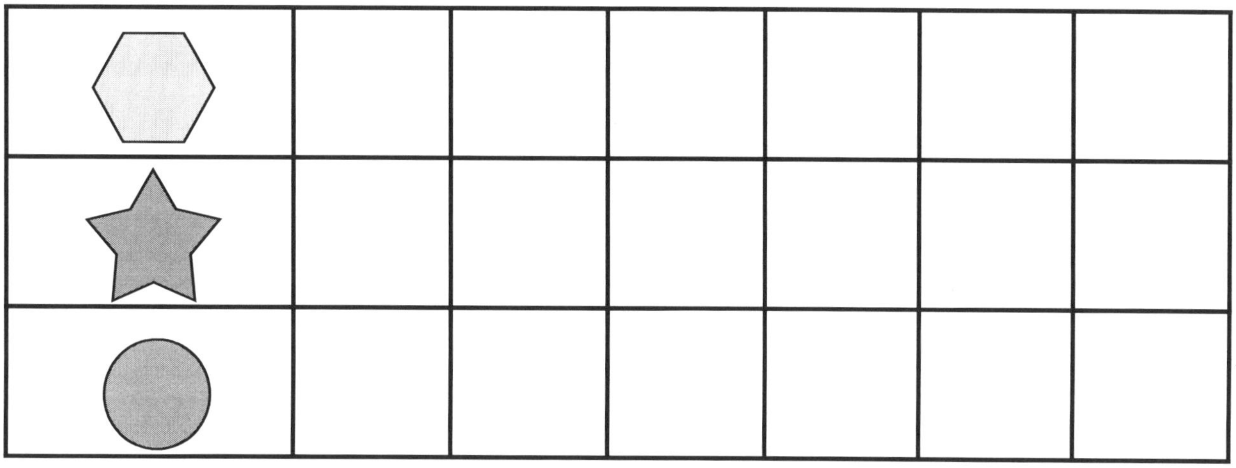

1. Which shape appeared the <u>most</u>?

2. Which shape appeared the <u>least</u>?

**BRAIN STRETCH:** Count the tally marks.

How many?

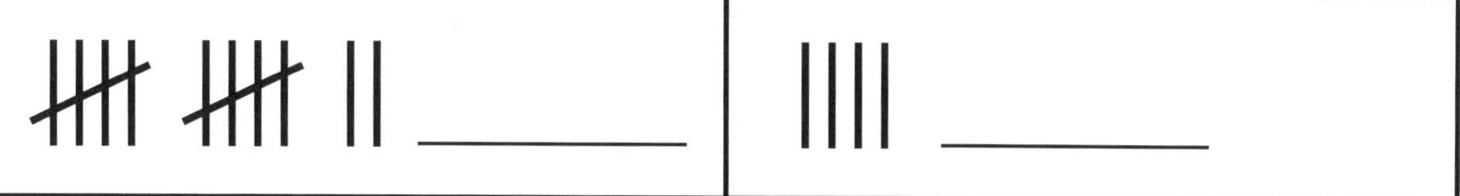

# EXPLORING BAR GRAPHS

Count the pictures and complete the bar graph.

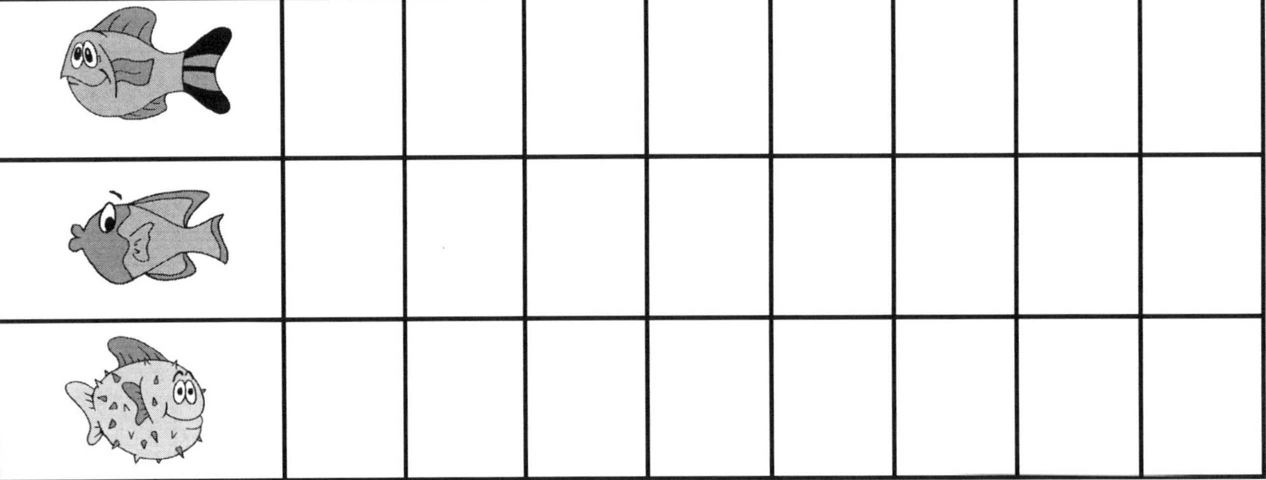

1. How many? 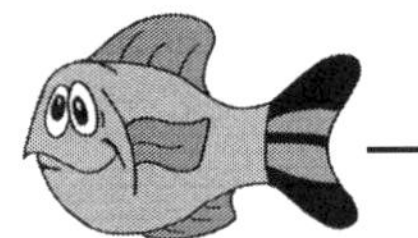 ______ 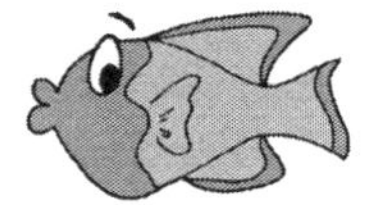 ______  ______

2. Which fish appeared the most?   

3. Which fish appeared the least?   

4. How many fish altogether?

# EXPLORING BAR GRAPHS

Count the pictures and complete the bar graph.

## BIRD GRAPH

1. How many?   ______  ______

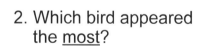

2. Which bird appeared the <u>most</u>?

3. Which bird appeared the <u>least</u>?

4. How many birds altogether?

# READING BAR GRAPHS

Mrs. Poulos' class made a bar graph to show the results of their favourite ball survey. Use the bar graph to answer the questions.

## FAVOURITE BALL

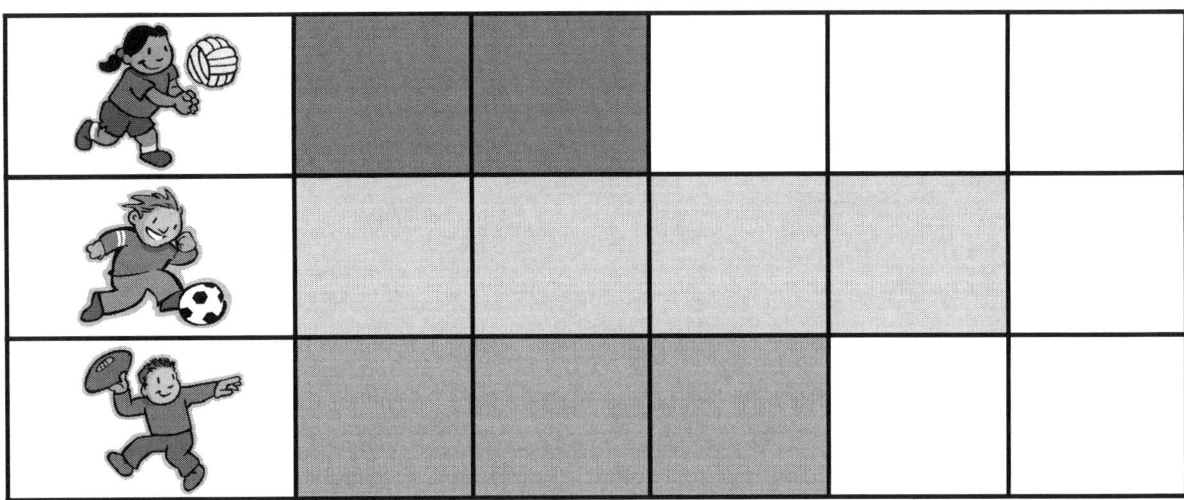

**BRAIN STRETCH:** Answer the questions.

1. How many votes?  ______ ______  ______

2. Circle the <u>most</u> popular ball.

3. Circle the <u>least</u> popular ball.

**BRAIN STRETCH:** Count the tally marks.

How many?

|  |  |
|---|---|

# READING BAR GRAPHS

Ms. Smith's class made a bar graph to show the results of their favourite winter activity survey. Use the bar graph to answer the questions.

## FAVOURITE WINTER ACTIVITY

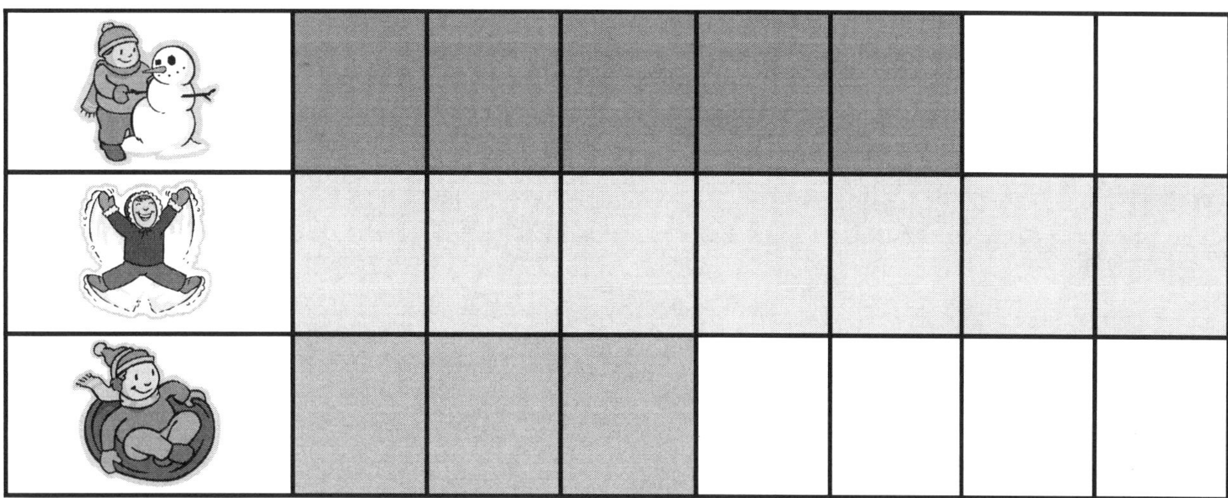

1. How many votes?  ______  ______  ______

2. Circle the most popular winter activity.   

3. Circle the least popular winter activity.   

**BRAIN STRETCH:** Count the tally marks.

How many?

 ______

 ______

# READING BAR GRAPHS

Ms. Schwartz's class made a bar graph to display the results of a survey on what instruments students would like to learn to play.

## INSTRUMENTS STUDENTS WOULD LIKE TO LEARN TO PLAY

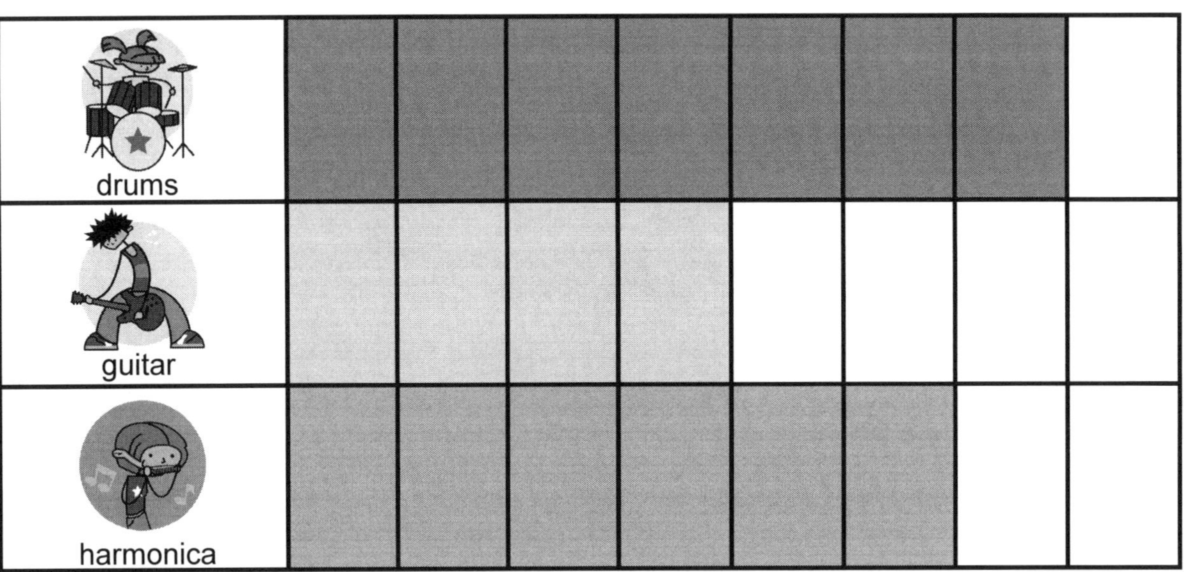


**BRAIN STRETCH:** Answer the questions.

1. How many votes?  ______  ______  ______

| | | | |
|---|---|---|---|
| **2.** | What kind of instrument do most students want to learn to play? | **3.** | What instrument do the least number of students want to learn to play? |
| **4.** | How many students want to learn to play the drums and the guitar? | **5.** | How many fewer students want to learn to play the guitar than the harmonica? |

# READING BAR GRAPHS

Mrs Simpson took a survey of students' favourite places to visit.

## FAVOURITE PLACES TO VISIT

Places

Library

Mall

Circus

Park

Zoo

Number of Votes

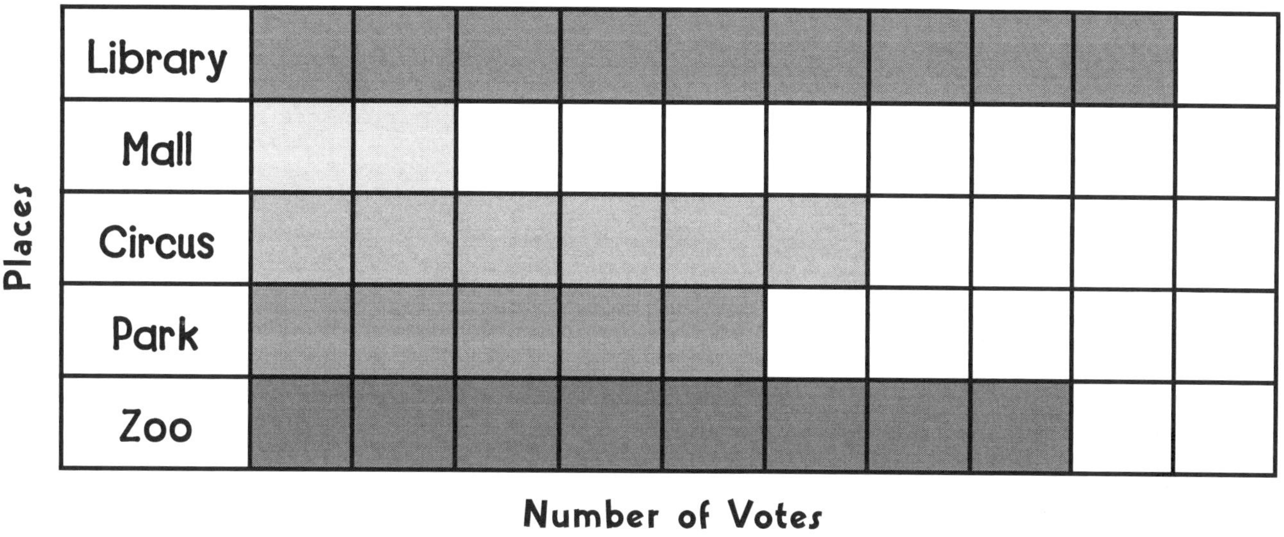

1. List the favourite places to visit in order from the least number of votes to the most.

a. ____________________ b. ____________________

e. ____________________

c. ____________________ d. ____________________

| | | | |
|---|---|---|---|
| **2.** | How many fewer votes are there for the mall than zoo? | **3.** | How many votes does the zoo have? |
| **4.** | What is the least popular place to visit? | **5.** | What is the most popular place to visit? |

# READING BAR GRAPHS

Here is a bar graph to show what type of lunch students like to eat in Mr. Moreno's class.

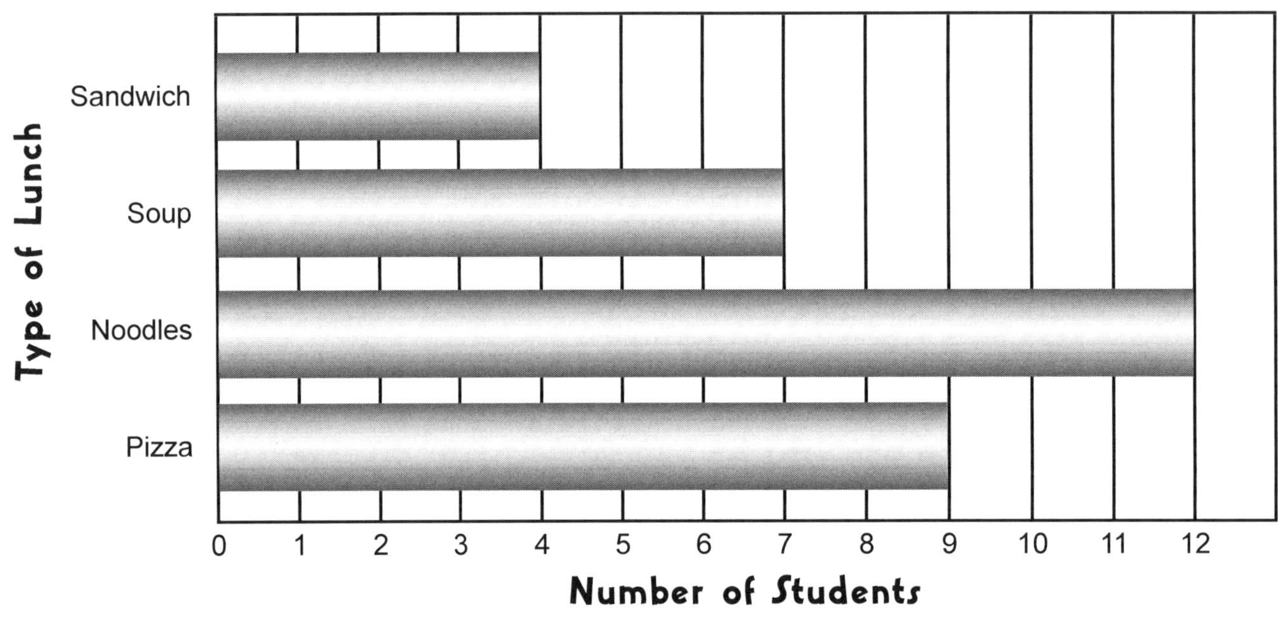


**BRAIN STRETCH:** Answer the questions.

| | | | |
|---|---|---|---|
| **1.** | Which type of lunch is the least popular? | **2.** | Which type of lunch is the most popular? |
| **3.** | How many students chose pizza? | **4.** | How many more votes did noodles get than sandwich? |
| **5.** | How many fewer votes did pizza get than noodles? | **6.** | Which type of lunch got 7 votes? |

# READING BAR GRAPHS

Here is a bar graph to show what kind of presents students like to receive in Ms. Regal's class.

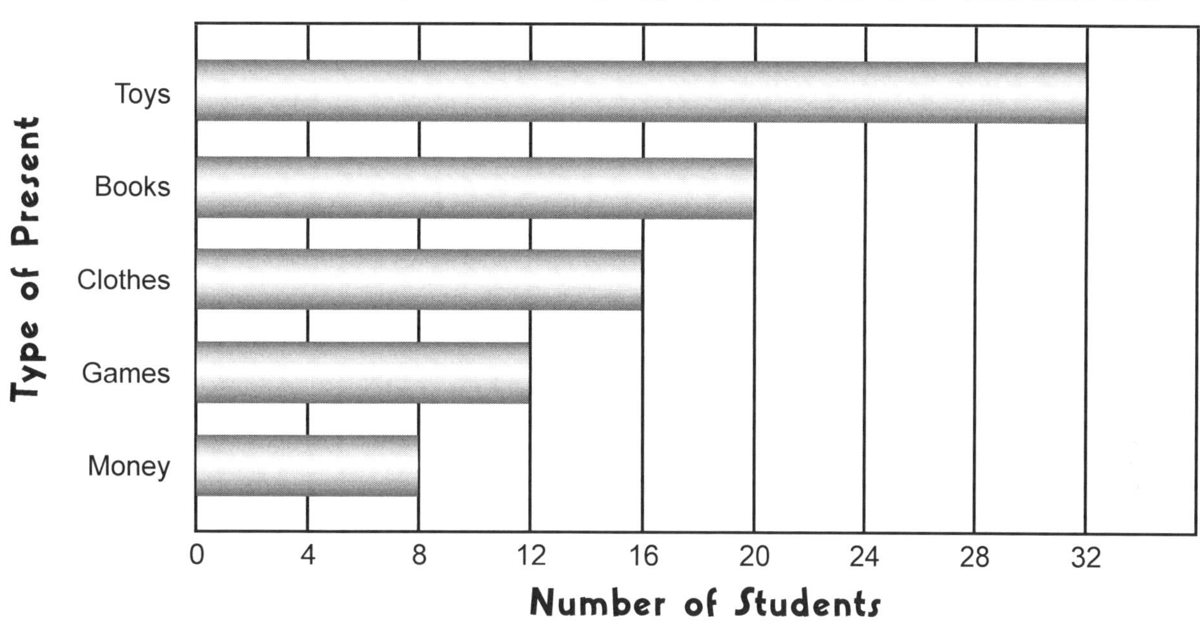


**BRAIN STRETCH:** Answer the questions.

| | | | |
|---|---|---|---|
| **1.** | Which type of present is the least popular? | **2.** | Which type of present is the most popular? |
| **3.** | How many students chose games? | **4.** | How many more votes did books get than money? |
| **5.** | How many fewer votes did games get than clothes? | **6.** | Which type of present got 16 votes? |

# READING BAR GRAPHS

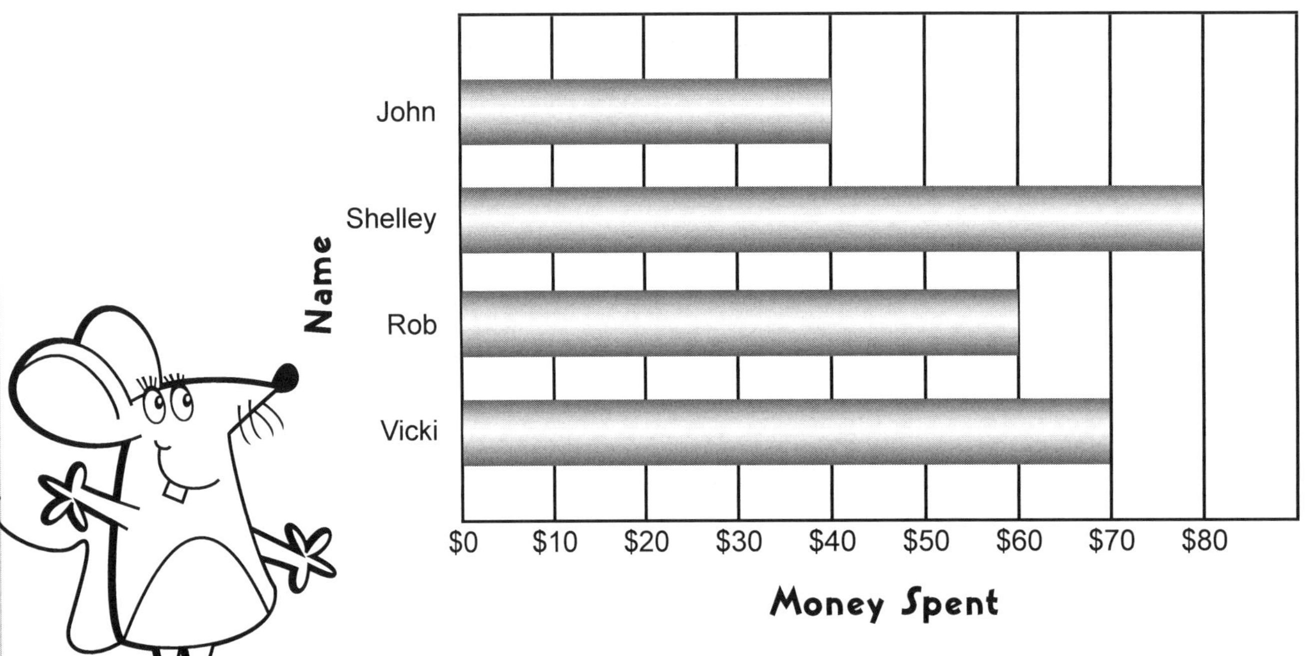


**BRAIN STRETCH:** Answer the questions.

| | | | |
|---|---|---|---|
| **1.** | Who spent the most money? | **2.** | How much more money did Shelley spend than John? |
| **3.** | Who spent $70? | **4.** | What is the difference between what Rob and Vicki spent? |
| **5.** | How much money did Vicki and Shelley spend altogether? | **6.** | Who spent $60? |

# READING BAR GRAPHS

The graph shows how many of each kind of fruit tree are in the Poulos' family orchard.

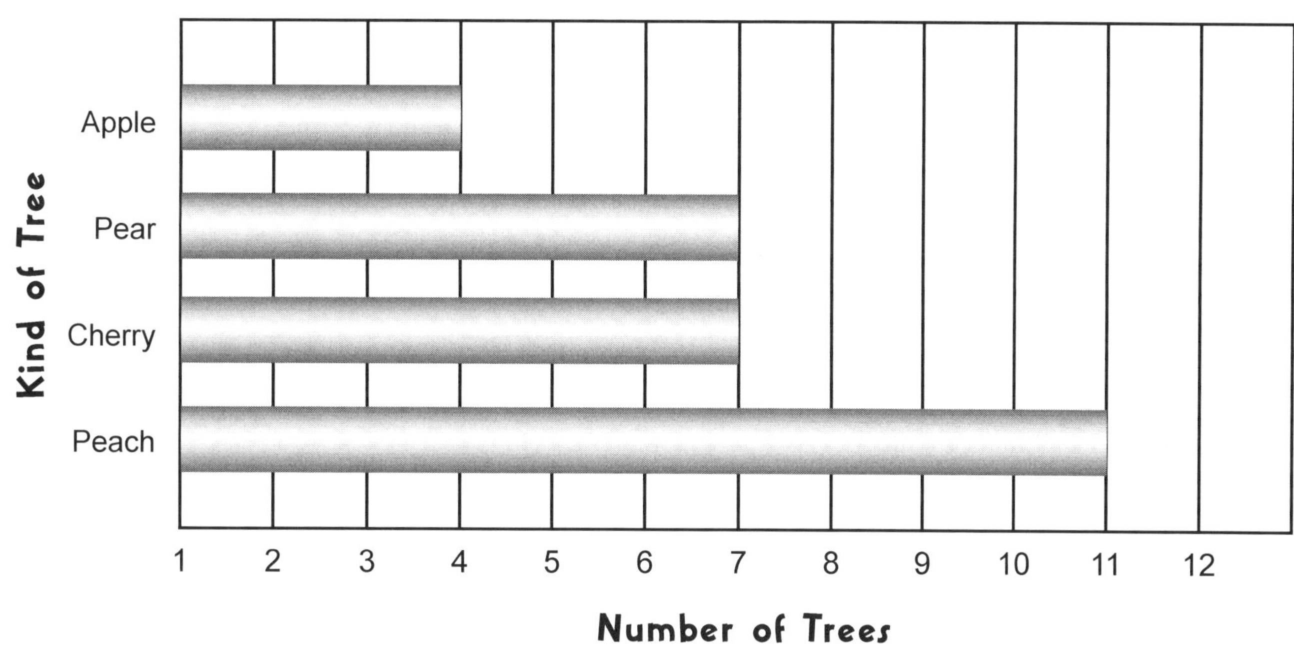


**BRAIN STRETCH:** Answer the questions.

| | | | |
|---|---|---|---|
| **1.** | Which kind of tree appeared the most? | **2.** | Which kind of tree appeared the least? |
| **3.** | Which kinds of trees appeared equally? | **4.** | How many fewer apple trees were there than cherry? |
| **5.** | How many pear trees were in the orchard? | **6.** | How many trees were in the orchard altogether? |

# READING BAR GRAPHS

Here are the results of a favourite fruit survey.

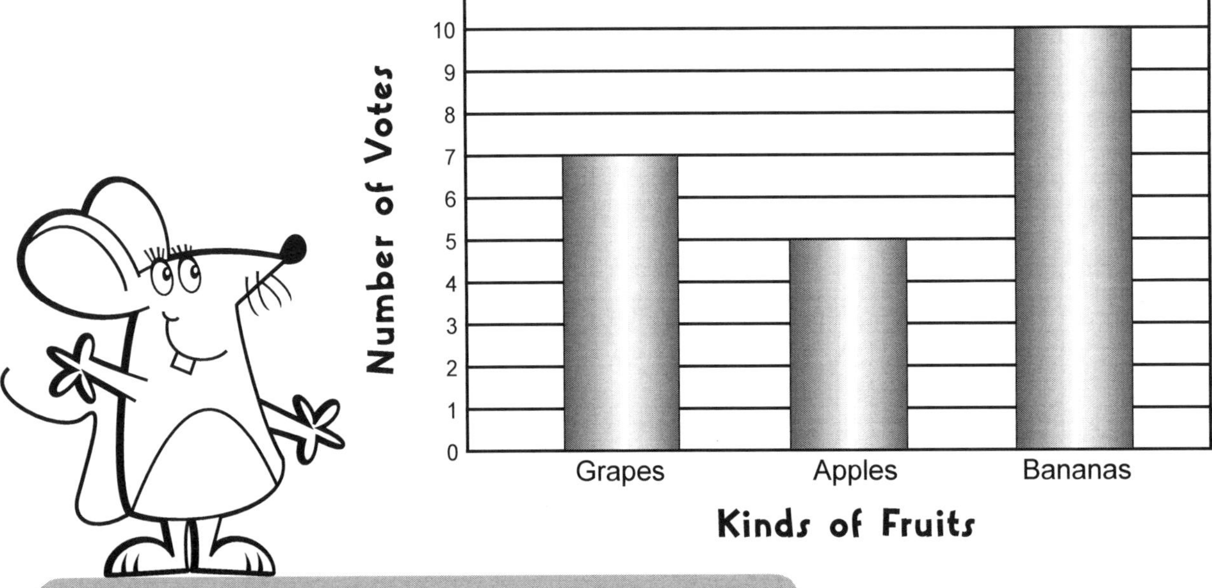


**BRAIN STRETCH:** Answer the questions.

1. List the fruits in order from the most number of votes to the least.

a. ______________________ b. ______________________

c. ______________________

| | | | |
|---|---|---|---|
| **2.** | How many more votes are there for grapes than for apples? | **3.** | How many votes does banana have? |
| **4.** | What is the least popular fruit? | **5.** | How many people voted in the survey altogether? |

# READING BAR GRAPHS

Here is a bar graph about students' collections.

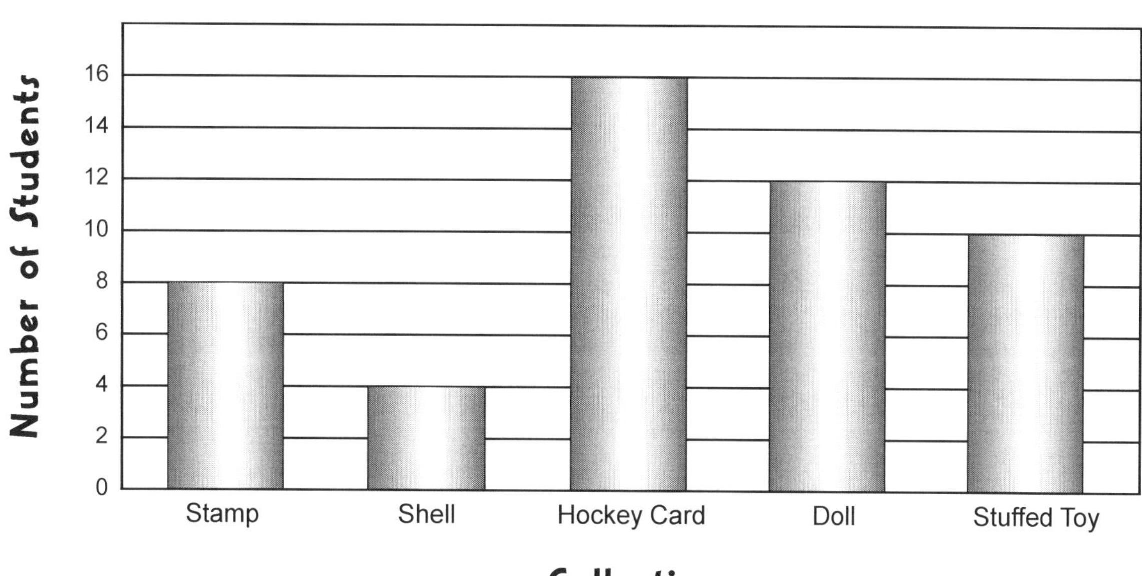


**BRAIN STRETCH:** Answer the questions.

1. List the collections in order from the collection students have the most to the collection students have the least.

a. ____________________ b. ____________________

c. ____________________ d. ____________________ e. ____________________

| | | | |
|---|---|---|---|
| **2.** | What kind of collection do most students have? | **3.** | What kind of collection do the least number of students have? |
| **4.** | How many students have either a shell collection or doll collection? | **5.** | How many students have a stuffed toy collection? |

# READING BAR GRAPHS

Here is a bar graph to show the number of glasses of lemonade sold at a lemonade stand.

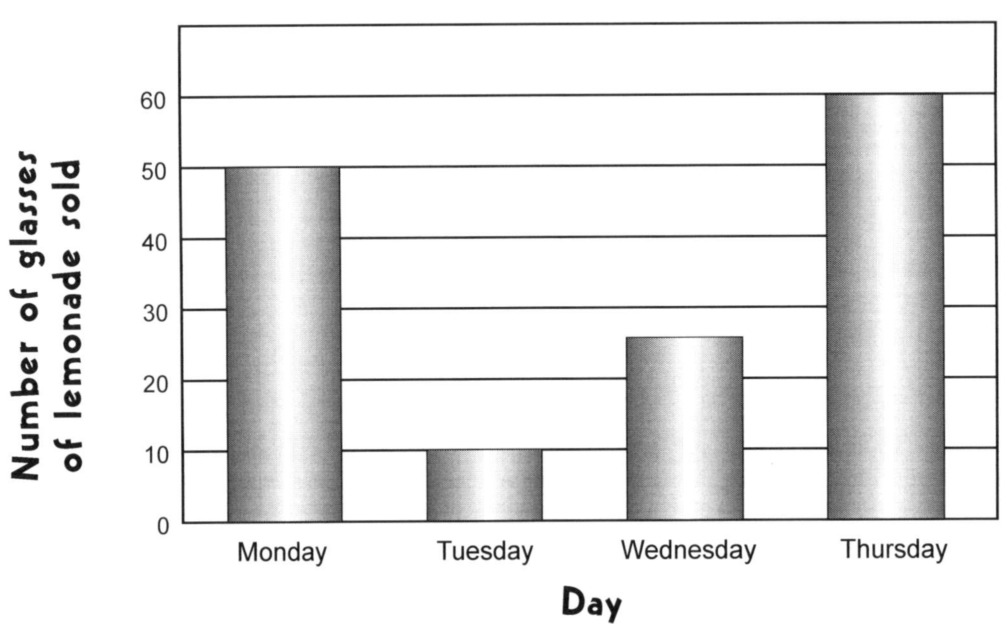


**BRAIN STRETCH:** Answer the questions.

| | | | |
|---|---|---|---|
| **1.** | On what day did Sophie sell the most lemonade? | **2.** | How many glasses of lemonade did Sophie sell on Tuesday and Wednesday? |
| **3.** | On what day did Sophie sell 25 glasses of lemonade? | **4.** | On what day did Sophie sell the least lemonade? |
| **5.** | How many fewer glasses of lemonade did Sophie sell on Monday than on Thursday? | **6.** | On what day did Sophie sell 50 glasses of lemonade? |

# READING BAR GRAPHS

The Demaat family went apple picking. Here is a bar graph to show how many apples each family member picked.

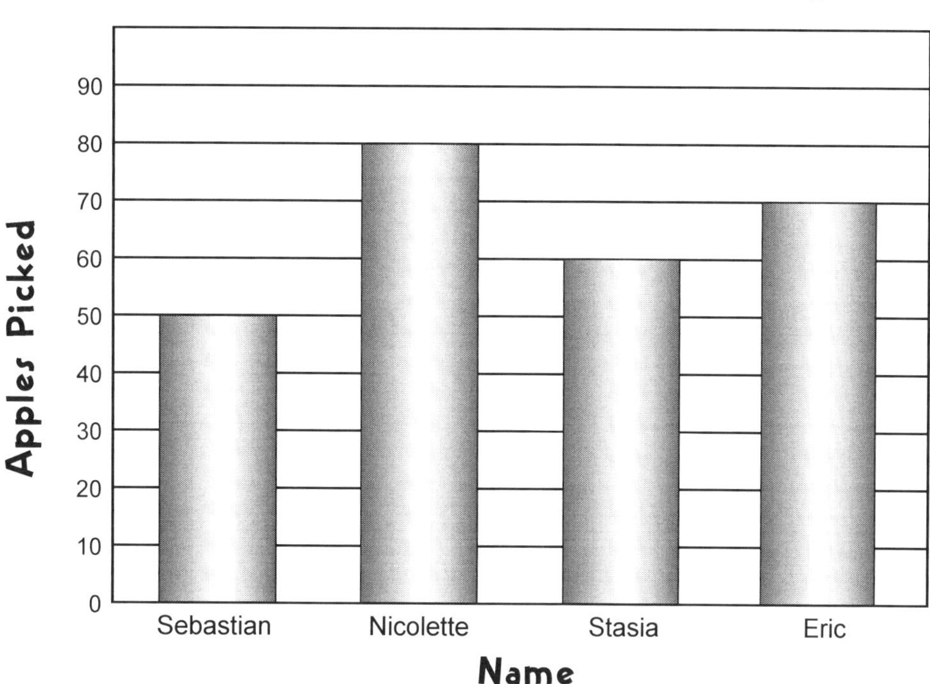


**BRAIN STRETCH:** Answer the questions.

| | | | |
|---|---|---|---|
| **1.** | Who picked the most number of apples? | **2.** | Who picked the least number of apples? |
| **3.** | How many more apples did Nicolette pick than Sebastian? | **4.** | How many apples did Stasia and Eric pick altogether? |
| **5.** | How many fewer apples did Stasia pick than Eric? | **6.** | How many apples were picked altogether? |

# READING BAR GRAPHS

Katherine and Alexander planted a garden. Here is a bar graph to show how many of each flower was planted.

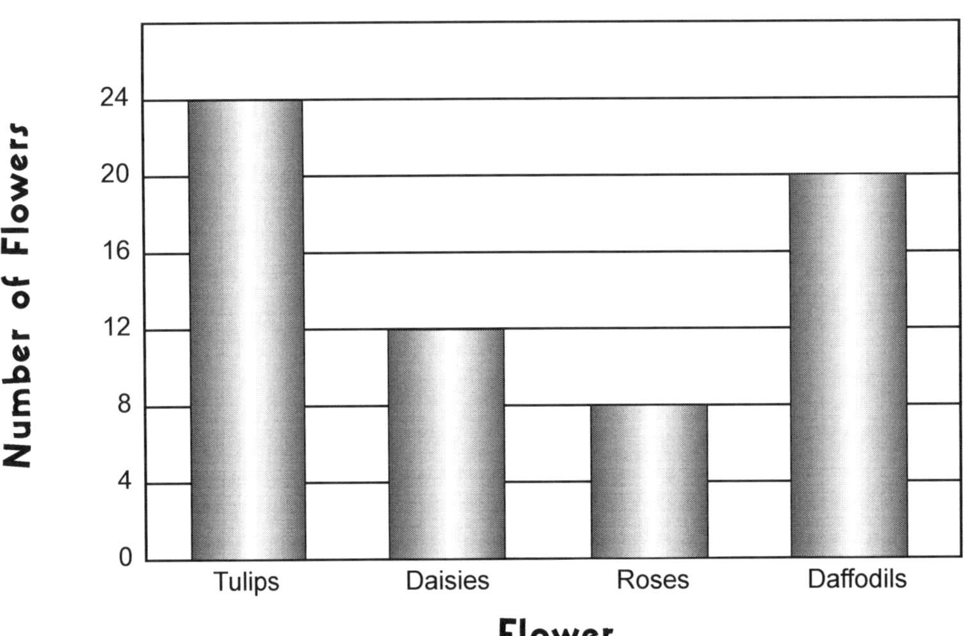


**BRAIN STRETCH:** Answer the questions.

| | | | |
|---|---|---|---|
| **1.** | Which kind of flower was planted the most? | **2.** | Which kind of flower was planted the least? |
| **3.** | How many more daisies than roses were planted? | **4.** | How many fewer daffodils were planted than tulips? |
| **5.** | How many roses were there? | **6.** | How many flowers were in the flower garden altogether? |

# READING BAR GRAPHS

David did a survey on the number of family members in his friends' families. Here is a bar graph to show the results.

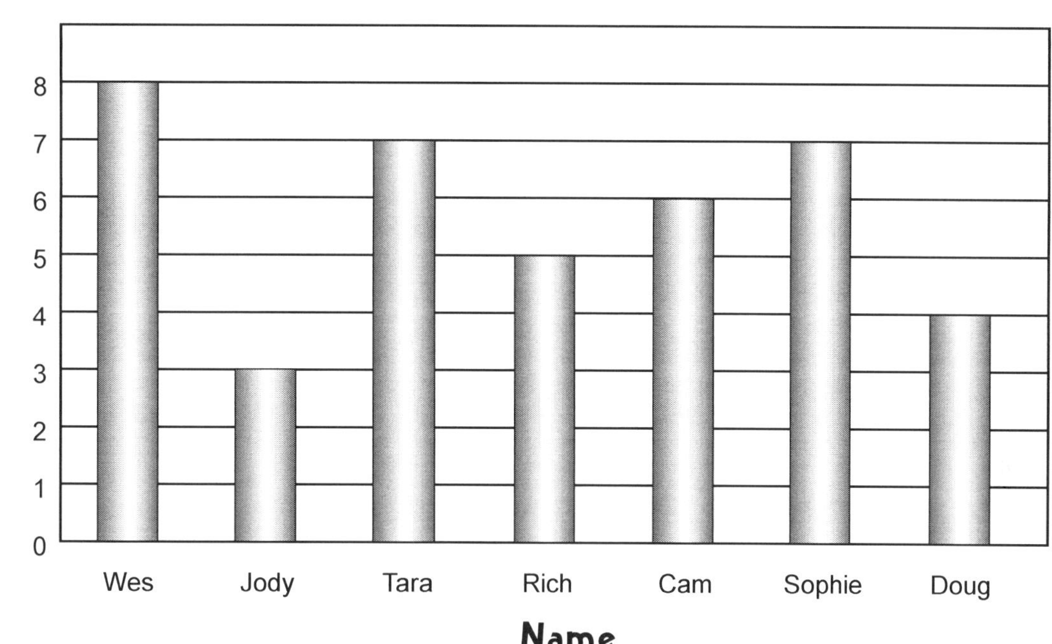


## BRAIN STRETCH: Answer the questions.

| | | | |
|---|---|---|---|
| **1.** | What would be a good title for this graph? | **2.** | Which 2 children have the same number of family members? |
| **3.** | How many fewer family members does Jody have than Rich? | **4.** | How many more family members does Sophie have than Doug |
| **5.** | How many children participated in this survey? | **6.** | Who has 5 family members? |

# EXPLORING GRAPHS

Use the data from the picture graph to make a bar graph.

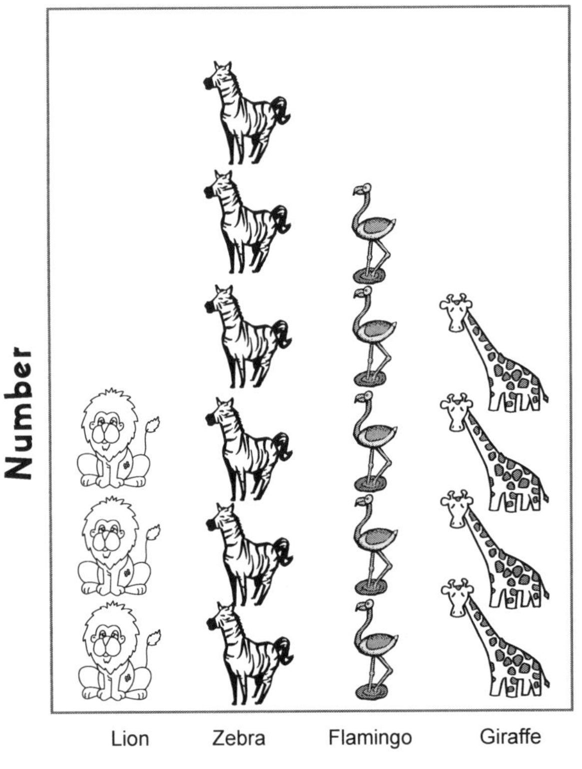


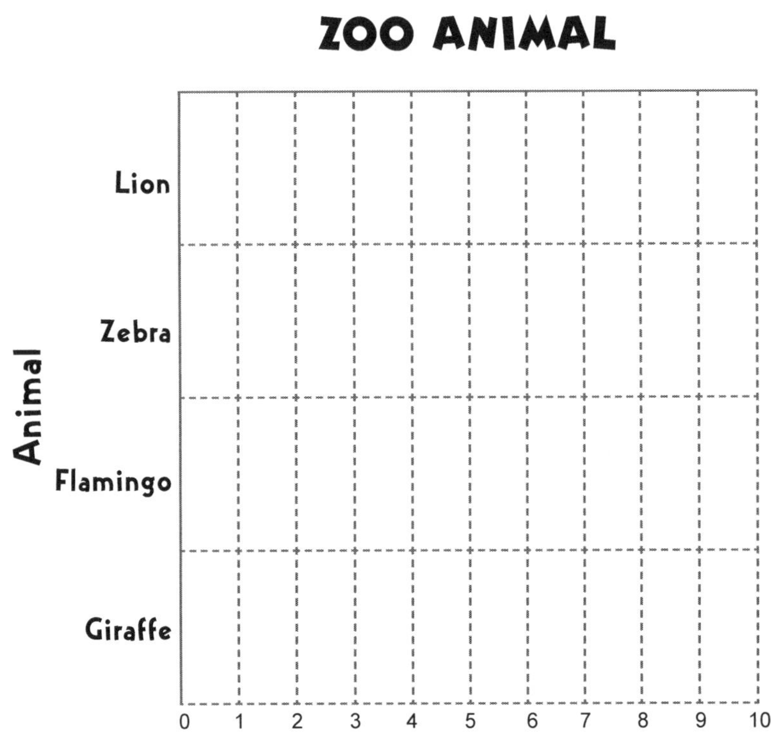


**BRAIN STRETCH:** Answer the questions.

1. How many people answered the survey? ______________________________

2. What is the most popular zoo animal? ______________________________

3. What is the least popular zoo animal? ______________________________

4. List the zoo animals in order from the zoo animal with the most votes to the zoo animal with the fewest votes.

   a. __________________ b. __________________

   c. __________________ d. __________________

# EXPLORING GRAPHS

Use the data from the picture graph to make a bar graph.

**FAVOURITE SNACK**

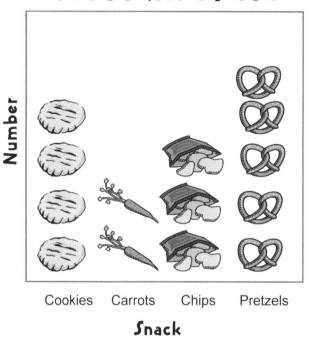

Number

Cookies Carrots Chips Pretzels

Snack

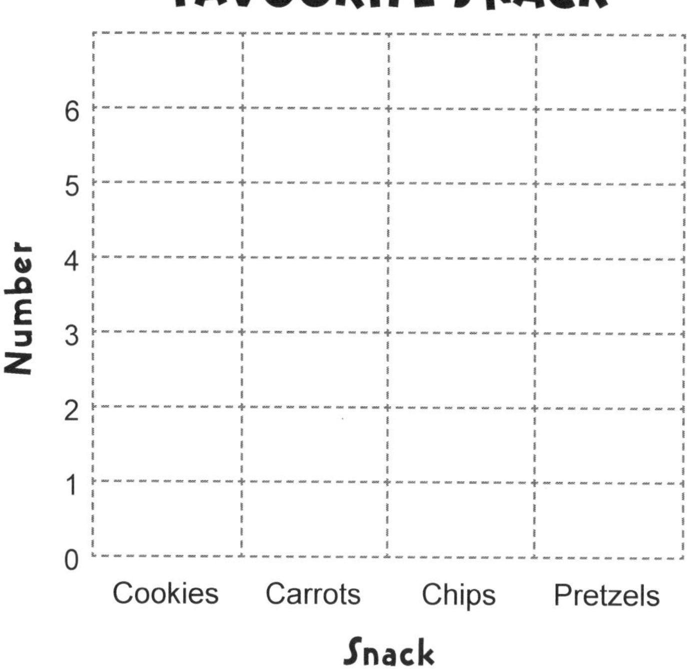


**BRAIN STRETCH:** Answer the questions.

| | | | |
|---|---|---|---|
| **1.** | Did more children choose chips or carrots? | **2.** | If 2 more children chose cookies how many total people would have chosen cookies? |
| **3.** | How many children voted in this survey? | **4.** | Which snack had 3 votes? |
| **5.** | What is the most popular snack? | **6.** | List the snacks from the snack with least votes to the snack with most the votes. |

# CONSTRUCTING BAR GRAPHS

Use data from the chart to complete the bar graph. Make sure you label!

| Vegetable | Number of Votes |
|---|---|
| Carrots | 6 |
| Broccoli | 4 |
| Peas | 8 |
| Potatoes | 3 |

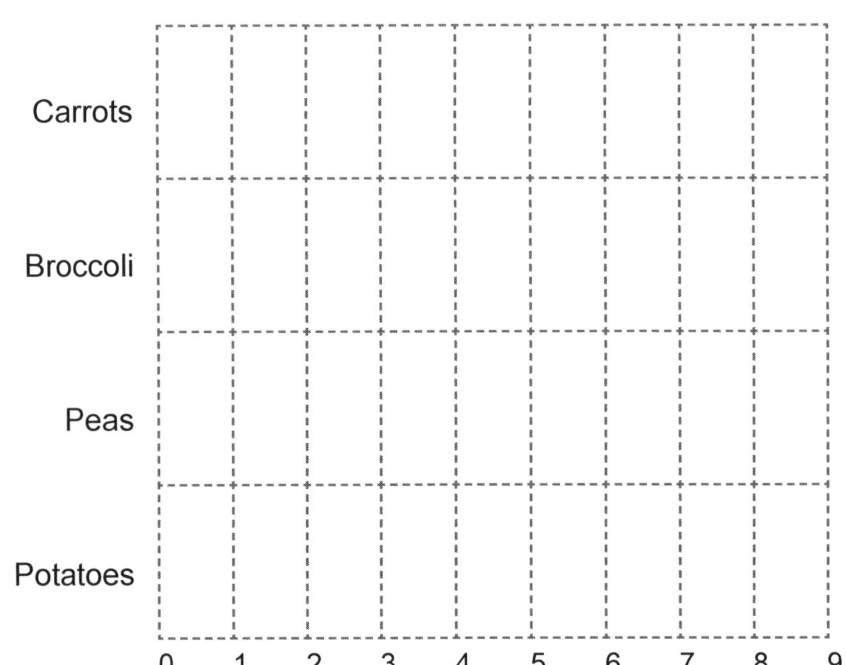


**BRAIN STRETCH:** Answer the questions.

**2.** List the vegetables in order from the least votes to the most.

**a.** ______________________ **b.** ______________________

**c.** ______________________ **d.** ______________________

| | | | |
|---|---|---|---|
| **2.** | How many more votes are there for peas than carrots? | **3.** | How many votes does broccoli have? |
| **4.** | Which vegetable have the least votes? | **5.** | What is the most popular vegetable? |

# EXPLORING BAR GRAPHS

Josh did a survey of the colour of cars found at the Super Mart parking lot. Complete the bar graph using the data collected.

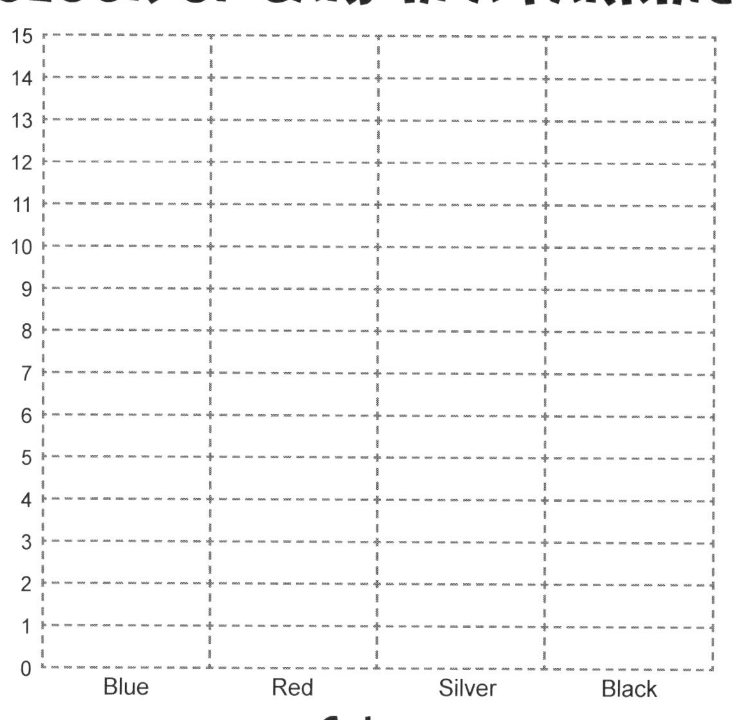

| Colour of Car | Number |
|---|---|
| Blue | 12 |
| Silver | 5 |
| Red | 8 |
| Black | 14 |

**BRAIN STRETCH:** Answer the questions.

| | | | |
|---|---|---|---|
| **1.** | How many fewer silver cars are there than black cars? | **2.** | Which colour has the least amount of cars in the parking lot? |
| **3.** | How many fewer red cars are there than blue? | **4.** | How many cars were there red and blue altogether? |
| **5.** | How many silver cars are there? | **6.** | How many more cars were needed so there would be 12 red cars? |

# EXPLORING BAR GRAPHS

Use data from the chart to complete the bar graph. Make sure you label!

| Season | Number of Votes |
|---|---|
| Spring | 13 |
| Summer | 15 |
| Autumn | 10 |
| Winter | 14 |

15
14
13
12
11
10
9
8
7
6
5
4
3
2
1
0

Spring Summer Autumn Winter

**BRAIN STRETCH:** Answer the questions.

**1.** List the seasons in order from the most votes to the fewest votes.

**a.** ____________________ **b.** ____________________

**c.** ____________________ **d.** ____________________

| | | | |
|---|---|---|---|
| **2.** | How many more people voted for winter than for spring? | **3.** | What season got 13 votes? |
| **4.** | How many people voted for summer and autumn altogether? | **5.** | What is the most popular season? |

# EXPLORING GRAPHS

This data table shows the results of hair colour survey taken by Mr. Turnbull's class. Construct a graph to display the data. Make sure you label your graph!

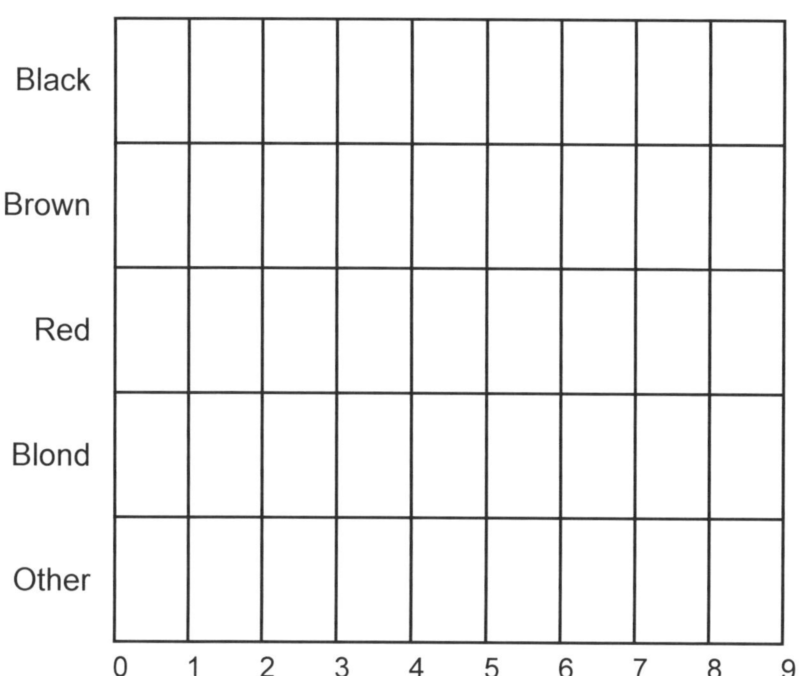


| Colour of Hair | Number |
|---|---|
| Black | 4 |
| Brown | 6 |
| Red | 9 |
| Blond | 5 |
| Other | 5 |

**BRAIN STRETCH:** Write about what you know by looking at the graph.

# EXPLORING BAR GRAPHS

Use data from the chart to make a bar graph. Make sure you label!

| Season | Number of Votes |
|---|---|
| Spring | 6 |
| Summer | 10 |
| Autumn | 5 |
| Winter | 15 |

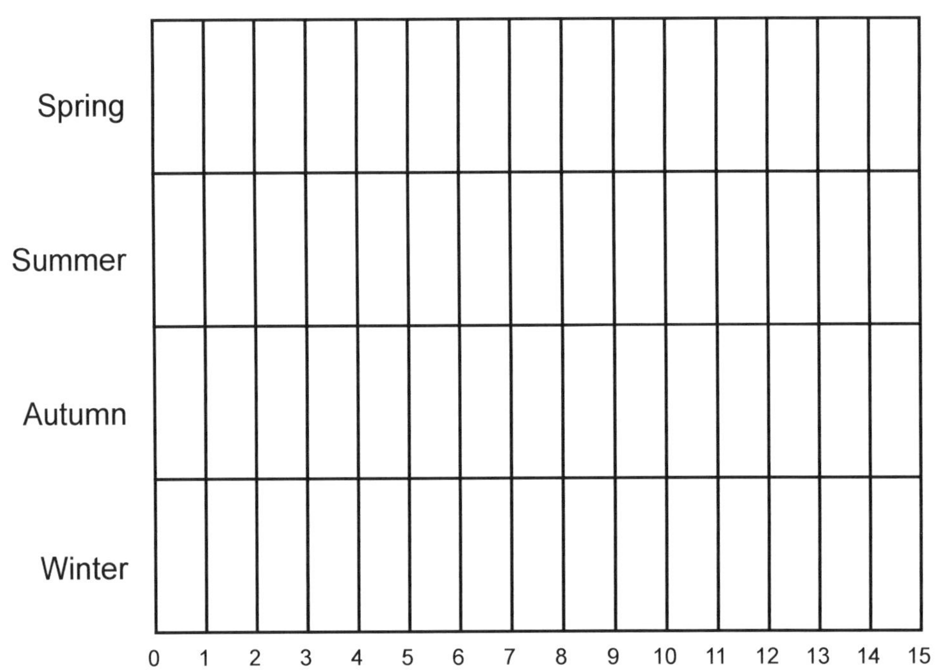


**BRAIN STRETCH:** Answer the questions.

1. List the seasons in order from the most votes to the fewest votes.

______________________________________________

2. How many more people chose winter than chose spring? ____________

3. How many people chose summer and winter altogether? ____________

4. What was the least popular season? ____________

5. What was the most popular season? ____________

6. How many fewer people chose autumn than summer? ____________

# EXPLORING BAR GRAPHS

The two grade 3 classes made a chart to record their data about some of their favourite pizza toppings.

| Pepperoni | Cheese | Tomatoes | Anchovies | Mushrooms |
|---|---|---|---|---|
| 25 | 35 | 15 | 10 | 15 |

Create a bar graph to show the information on the chart. Make sure you put on labels and a title.

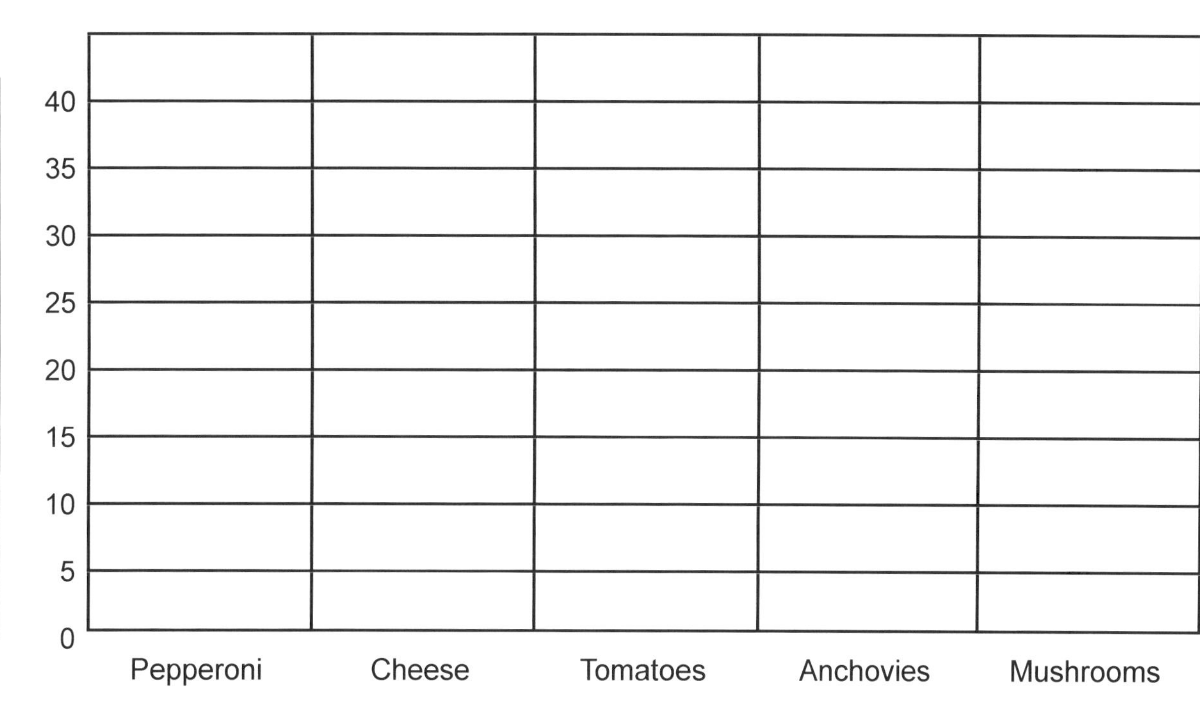


**BRAIN STRETCH:** What do you know by looking at the graph?

# EXPLORING BAR GRAPHS

The two grade 3 classes conducted a survey on students' favourite subjects. Here are the results. Create a bar graph to show the information. Make sure you label your graph!

| Reading | Art | Math | Science | Music |
|---|---|---|---|---|
| 35 | 25 | 15 | 10 | 20 |

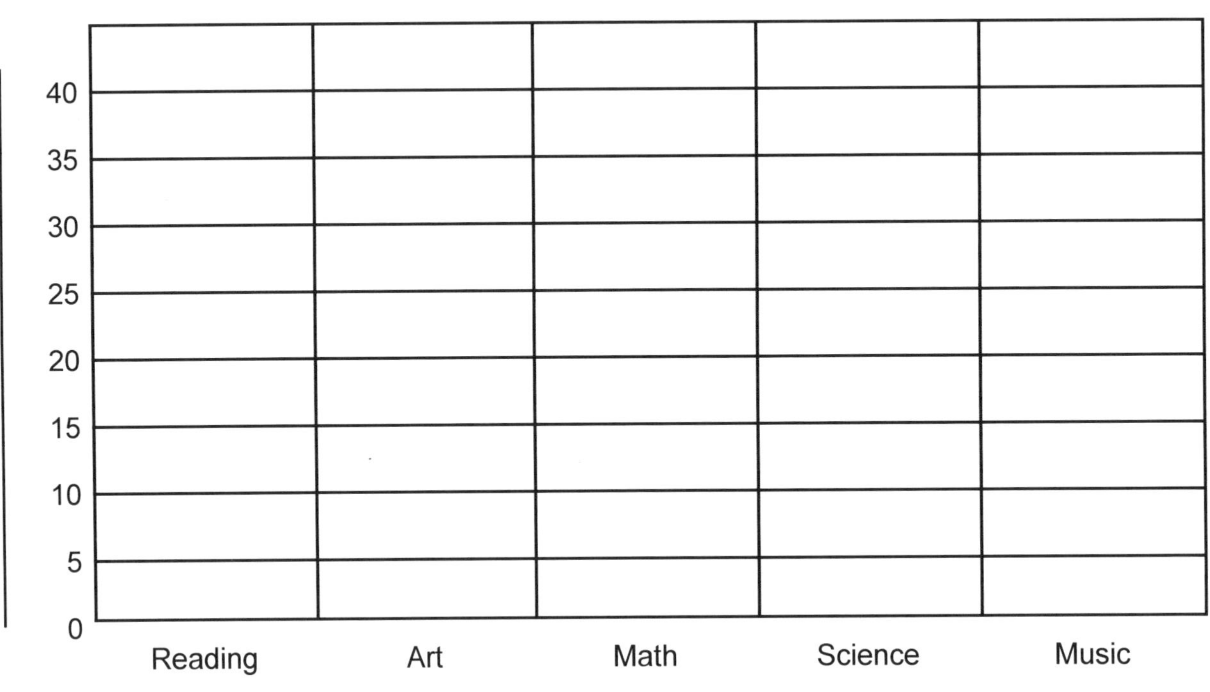


**BRAIN STRETCH:** Write about what you know by looking at the graph.

# READING A CALENDAR

| February | | | | | | |
|---|---|---|---|---|---|---|
| S | M | T | W | T | F | S |
| | | | | | 1 | 2 |
| 3 | 4 | 5 | 6 | 7 | 8 | 9 |
| 10 | 11 | 12 | 13 | 14 | 15 | 16 |
| 17 | 18 | 19 | 20 | 21 | 22 | 23 |
| 24 | 25 | 26 | 27 | | | |

**BRAIN STRETCH:** Use the calendar to answer the questions.

| | |
|---|---|
| 1. What day of the week is February 18?<br><br>____________________ | 2. What is the date of the first Tuesday?<br><br>____________________ |
| 3. How many Saturdays are there?<br><br>____________________ | 4. What is the date of the second Thursday?<br><br>____________________ |
| 5. On what day of the week will the next month begin?<br><br>____________________ | 6. What day of the week is February 15?<br><br>____________________ |
| 7. What is the date of the first Friday?<br><br>____________________ | 8. How many Sundays are there?<br><br>____________________ |
| 9. What is the date of the third Monday?<br><br>____________________ | 10. What is the date of the first Wednesday?<br><br>____________________ |

# READING A CALENDAR

| March | | | | | | |
|---|---|---|---|---|---|---|
| S | M | T | W | T | F | S |
| | | | | | | 1 |
| 2 | 3 | 4 | 5 | 6 | 7 | 8 |
| 9 | 10 | 11 | 12 | 13 | 14 | 15 |
| 16 | 17 | 18 | 19 | 20 | 21 | 22 |
| 23 | 24 | 25 | 26 | 27 | 28 | 29 |
| 30 | 31 | | | | | |

**BRAIN STRETCH:** Use the calendar to answer the questions.

| | |
|---|---|
| 1. On what day of the week will the next month begin? ____________ | 2. How many Saturdays are there? ____________ |
| 3. What is the date of the second Wednesday? ____________ | 4. What is the date of the first Sunday? ____________ |
| 5. What day of the week is March 29? ____________ | 6. What is the date of the third Monday? ____________ |
| 7. What day of the week is March 17? ____________ | 8. How many Thursdays are there? ____________ |
| 9. What is the date of the first Tuesday? ____________ | 10. How many Fridays are there? ____________ |

# READING A CALENDAR

| August | | | | | | |
|---|---|---|---|---|---|---|
| S | M | T | W | T | F | S |
| | | | | | 1 | 2 |
| 3 | 4 | 5 | 6 | 7 | 8 | 9 |
| 10 | 11 | 12 | 13 | 14 | 15 | 16 |
| 17 | 18 | 19 | 20 | 21 | 22 | 23 |
| 24 | 25 | 26 | 27 | 28 | 29 | 30 |
| 31 | | | | | | |

**BRAIN STRETCH:** Use the calendar to answer the questions.

| | |
|---|---|
| 1. What day of the week is August 11?<br><br>____________________ | 2. On what day of the week will the next month begin?<br><br>____________________ |
| 3. What is the date of the third Tuesday?<br><br>____________________ | 4. What is the date of the first Monday?<br><br>____________________ |
| 5. How many Sundays are there?<br><br>____________________ | 6. What day of the week is August 30?<br><br>____________________ |
| 7. What is the date of the first Thursday?<br><br>____________________ | 8. What is the date of the fifth Saturday?<br><br>____________________ |
| 9. How many Fridays are there?<br><br>____________________ | 10. How many Wednesdays are there?<br><br>____________________ |

# READING A CALENDAR

| January | | | | | | |
|---|---|---|---|---|---|---|
| S | M | T | W | T | F | S |
| | | 1 | 2 | 3 | 4 | 5 |
| 6 | 7 | 8 | 9 | 10 | 11 | 12 |
| 13 | 14 | 15 | 16 | 17 | 18 | 19 |
| 20 | 21 | 22 | 23 | 24 | 25 | 26 |
| 27 | 28 | 29 | 30 | 31 | | |

**BRAIN STRETCH:** Use the calendar to answer the questions.

| | |
|---|---|
| 1. What is the date of the first Saturday?<br><br>____________________ | 2. What day of the week is January 16?<br><br>____________________ |
| 3. How many Wednesdays are there?<br><br>____________________ | 4. What is the date of the third Friday?<br><br>____________________ |
| 5. On what day of the week will the next month begin?<br><br>____________________ | 6. How many Tuesdays are there?<br><br>____________________ |
| 7. What is the date of the first Monday?<br><br>____________________ | 8. What day of the week is January 3?<br><br>____________________ |
| 9. What is the date of the fourth Thursday?<br><br>____________________ | 10. What day of the week is January 21?<br><br>____________________ |

# READING VENN DIAGRAMS

Use the Venn Diagram to answer the questions about the clubs Elizabeth and Alan are involved with.

## WHAT SCHOOL CLUBS ARE YOU INVOLVED IN?

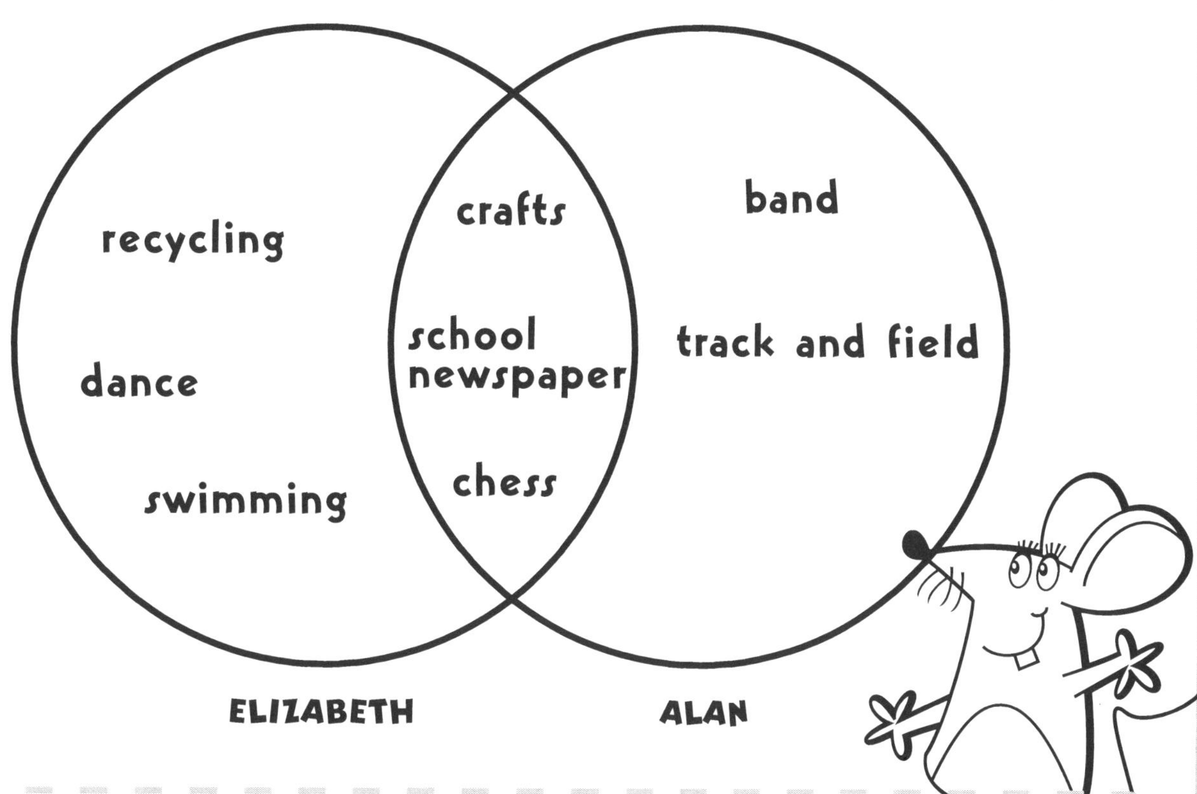


1. Which clubs is Elizabeth involved in? ______________________________

2. Which clubs is Alan involved in? ______________________________

3. Which clubs are both Elizabeth and Alan involved in? ______________

## BRAIN STRETCH:

If ☼ = 10 What is the value of ☼ ☼ ☼ ? __________

# READING VENN DIAGRAMS

Use the Venn Diagram to answer the questions about these students' favourite kind of milk.

1. Which students like white milk, but not chocolate milk? ____________________

2. Which students like chocolate milk, but not white milk? ____________________

3. Which students like both types of milk? ____________________

## BRAIN STRETCH:

If [cookie] = 2 What is the value of  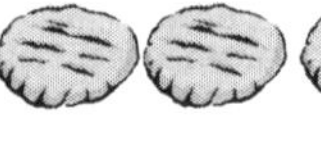  ? ____________

# READING VENN DIAGRAMS

Use the Venn Diagram to answer the questions about these students' favourite recess activities.

## FAVOURITE RECESS ACTIVITIES

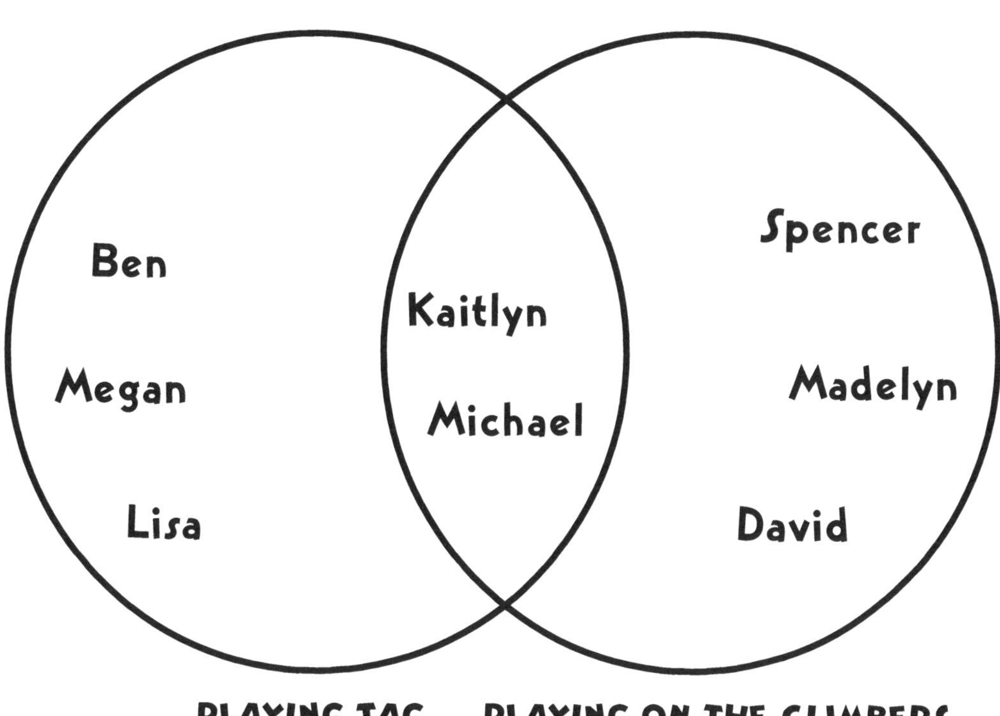


1. Which students like to play tag, but not play on the climbers? ______________

2. Which students like to play on the climbers, but not play tag? ______________

3. Which students like to do both activities? ______________________________

## BRAIN STRETCH:

If  = 4 What is the value of    ? ____________

# WRITING SURVEY QUESTIONS

Write a survey question you could ask for each of the following groups. Each survey question should have at least two choices.

### 1. STUDENTS IN YOUR GRADE

___

___

___

### 2. GROWN UPS

___

___

___

### 3. PEOPLE WHO LIKE MUSIC

___

___

___

### 4. PEOPLE FROM YOUR TOWN

___

___

___

# WRITING SURVEY QUESTIONS

Write a survey question you could ask for each of the following groups. Each survey question should have at least two choices.

## 1. YOUNG CHILDREN

## 2. TEENAGERS

## 3. ALL STUDENTS AT YOUR SCHOOL

## 4. STUDENTS WHO HAVE PETS

# SORTING FUN

Cut out and paste each item into the correct box.

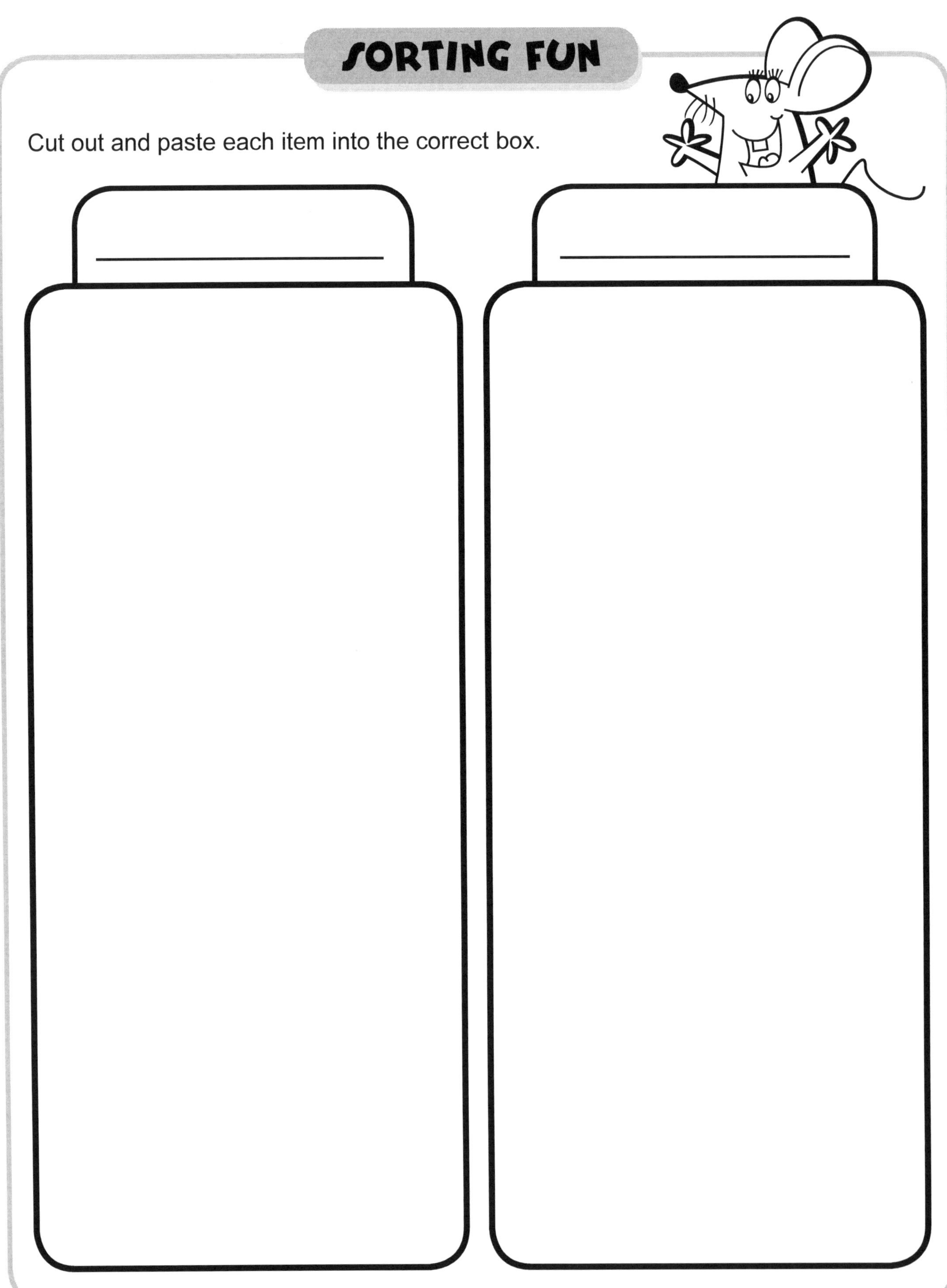

# MAKE A TALLY CHART

Use the information from the _______________ survey to complete the tally chart.

_______________________________ SURVEY

| | |
|---|---|
| | |
| | |
| | |
| | |
| | |
| | |
| | |
| | |
| | |

## TALLY CHART

| | Number of Votes |
|---|---|
| | |
| | |
| | |
| | |

## BRAIN STRETCH:

| 1. | | 2. | |
|---|---|---|---|

## ________________'s SURVEY AND RESULTS

Survey Question ______________________________________________

### TALLY CHART

| | Number of Votes |
|---|---|
| | |
| | |
| | |
| | |

______________ people took part in my survey.

### ______________________ BAR GRAPH

| | | | | | | | | |
|---|---|---|---|---|---|---|---|---|
| | | | | | | | | |
| | | | | | | | | |
| | | | | | | | | |
| | | | | | | | | |

**BRAIN STRETCH:**

On a separate piece of paper, write about what information you have learned from your line graph.

# EXPLORING BAR GRAPHS

Here is a data table to show ___________________________________

Complete the chart an and bar graph to show the information. Make sure to label!

| | Number of Votes | Number |
|---|---|---|
| | | |
| | | |
| | | |
| | | |
| | | |

_______________________________________

| | | | | | |
|---|---|---|---|---|---|
| 15 | | | | | |
| 14 | | | | | |
| 13 | | | | | |
| 12 | | | | | |
| 11 | | | | | |
| 10 | | | | | |
| 9 | | | | | |
| 8 | | | | | |
| 7 | | | | | |
| 6 | | | | | |
| 5 | | | | | |
| 4 | | | | | |
| 3 | | | | | |
| 2 | | | | | |
| 1 | | | | | |
| 0 | | | | | |

__________ __________ __________ __________ __________

_______________________________________

# READING VENN DIAGRAMS

Use the Venn Diagram to answer the questions about ______________________________

______________________________________________________________________

_______________________

_______________________ _______________________

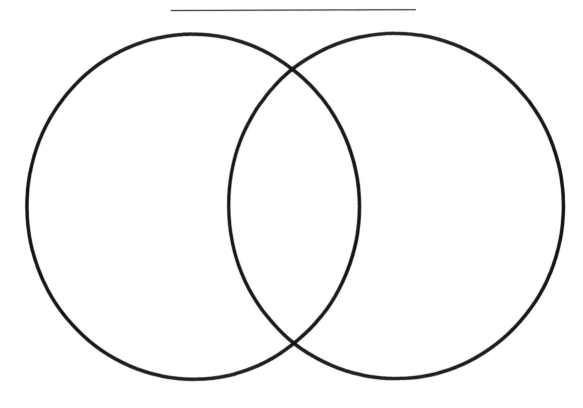

**BRAIN STRETCH:** Write what you know by looking at the Venn diagram.

______________________________________________________________________

______________________________________________________________________

______________________________________________________________________

______________________________________________________________________

# VENN DIAGRAM

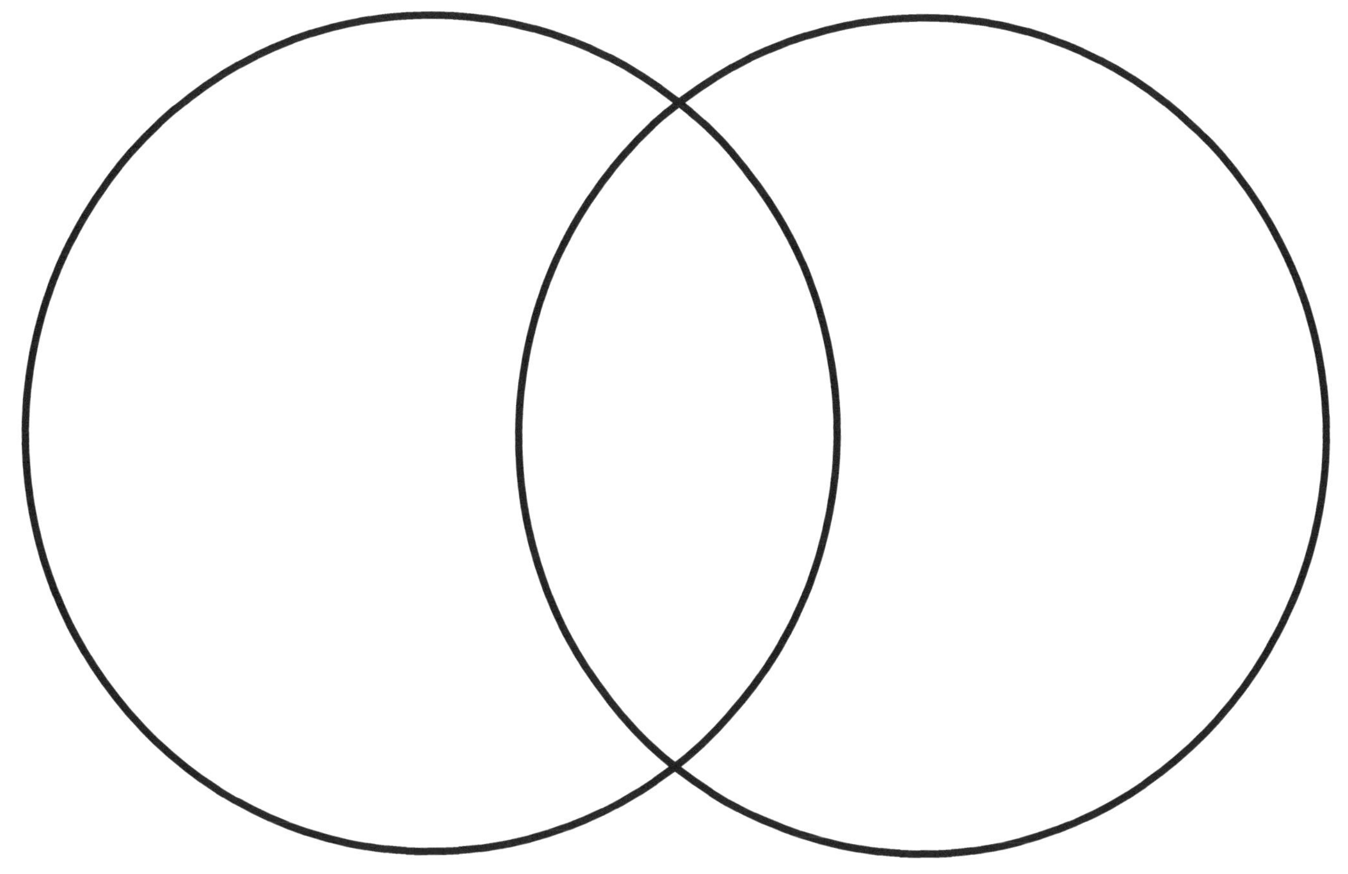

# DISPLAYING DATA

Data is collection of information usually gathered through observation, questioning or measurement.

## TALLY CHARTS

A tally chart shows data by counting by groups of five. Each line, or tally represents 1. Once you reach a group of 5 you start another group.

𝍸 = 5

| Colour | Tally | Number |
|---|---|---|
| Red | 𝍸 \|\| | |
| Blue | 𝍸 𝍸 \|\|\| | |
| Green | \|\|\|\| | |
| Purple | 𝍸 \|\|\| | |

## BAR GRAPHS

Bar graphs use horizontal or vertical bars that display data.

- Bar graphs are a good choice to display data if you want to compare data.

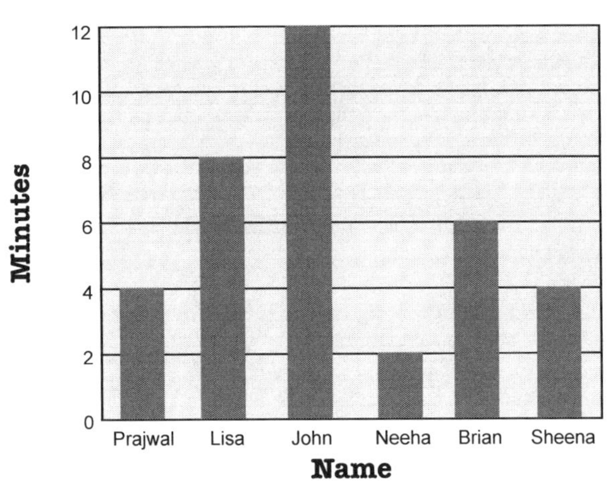


## PICTOGRAPHS

Pictographs use pictures or icons to show data and to compare information. Each picture or icon can represent more than one object. A key is used to what each picture represents.

- Pictographs are a good choice to display data if you want to compare data.

**Favourite Ice Cream Flavours**

| | |
|---|---|
| Strawberry | 🍦🍦🍦🍦🍦 |
| Chocolate Chip | 🍦🍦🍦 |
| Vanilla | 🍦🍦🍦🍦🍦🍦🍦 |
| Chocolate | 🍦🍦 |

 = 2 people

# WHOLE GROUP DATA SURVEY QUESTIONS

## WHAT IS YOUR FAVOURITE ________________ ?

- pet
- colour
- fruit
- home activity
- snack
- season
- winter activity
- summer activity
- sport
- recess activity
- pizza topping
- ice cream flavour
- dessert
- restaurant
- meal
- day of the week
- superhero
- author
- reading genre
- music genre
- school subject
- holiday
- cereal
- breakfast meal
- game
- coin
- cartoon
- lunch meal
- vegetable
- time of day
- farm animal
- dinner meal
- fast food
- movie genre
- candy
- month of the year
- weather
- community helper
- zoo animal
- transportation

## MAKE YOUR OWN QUESTIONS RELATED TO WHAT IS GOING ON IN YOUR CLASSROOM:

- What do you prefer?
- What do you like best?
- What is your estimation?
- What is your prediction?

## MORE IDEAS:

- How many family members are in your family?
- What colour hair do you have?
- How many teeth have you lost?
- Is your pet a mammal, reptile, bird, or fish?
- Where would you rather go on vacation?
- Would you rather live in the country or in the city?
- What colour eyes do you have?
- How many pets do you have?
- How many letters are there in your first name?
- How many hours of TV do you watch a day?
- What would you like to be when you grow up?
- What is your birthday month?

## QUESTIONS TO POSE ABOUT A GRAPH:

- What was the most?
- What was the least?
- How many more votes does ________ have than ________?
- How many fewer votes does ________ have than ________?
- How many altogether in two categories?
- How many people were asked in this survey?
- How many votes were there for ________?

# THINKING ABOUT DATA MANAGEMENT

How is data management used in every day life?

# GRADE 1 DATA MANAGEMENT

Student ____________________________________________

| Expectation | Level 1<br>Not Yet | Level 2<br>Developing | Level 3<br>Proficient | Level 4<br>Mastered |
|---|---|---|---|---|
| demonstrate the ability to organize objects into categories by sorting and classifying objects using one attribute (colour, size) , and by describing informal sorting experiences | | | | |
| read and describe primary data presented in concrete graphs and pictographs | | | | |
| collect and organize primary data | | | | |
| pose and answer questions about collected data | | | | |
| read primary data presented in concrete graphs and pictographs, and describe the data using comparative language | | | | |
| describe the likelihood that everyday events will occur, using mathematical language | | | | |

**Level 1** - Student rarely applies skills with several errors or omissions.

**Level 2** - Student sometimes applies skills with some errors or omissions.

**Level 3** - Student usually applies skills with few errors or omissions.

**Level 4** - Student consistently applies skills with almost no errors or omissions.

# GRADE 2 DATA MANAGEMENT

Student ______________________________________________

| Expectation | Level 1 Not Yet | Level 2 Developing | Level 3 Proficient | Level 4 Mastered |
|---|---|---|---|---|
| demonstrate an ability to organize objects into categories, by sorting and classifying objects using two attributes simultaneously | | | | |
| gather data to answer a question, using a simple survey with a limited number of responses | | | | |
| collect and organize primary data that is categorical or discrete and display the data using one-to-one correspondence in concrete graphs, pictographs, simple bar graphs, and other graphic organizers (e.g., tally charts, diagrams), with appropriate titles and labels and with labels ordered appropriately along horizontal axes, as needed | | | | |
| read primary data presented in concrete graphs, pictographs, simple bar graphs, and other graphic organizers(e.g., tally charts, diagrams), and describe the data using mathematical language | | | | |
| pose and answer questions about class generated data in concrete graphs, pictographs, simple bar graphs, and tally charts | | | | |
| distinguish between numbers that represent data values and numbers that represent the frequency of an event | | | | |
| demonstrate an understanding of data displayed in a graph, by comparing different parts of the data and by making statements about the data as a whole | | | | |

**Level 1** - Student rarely applies skills with several errors or omissions.

**Level 2** - Student sometimes applies skills with some errors or omissions.

**Level 3** - Student usually applies skills with few errors or omissions.

**Level 4** - Student consistently applies skills with almost no errors or omissions.

# GRADE 3 DATA MANAGEMENT

Student ___________________________________________

| Expectation | Level 1<br>Not Yet | Level 2<br>Developing | Level 3<br>Proficient | Level 4<br>Mastered |
|---|---|---|---|---|
| read, describe, and interpret primary data presented in charts and graphs, including vertical and horizontal bar graphs | | | | |
| sort and classify objects using two or more attributes simultaneously | | | | |
| read primary data presented in charts, tables, and graphs | | | | |
| collect data by conducting a simple survey about themselves, their environment, issues in their school or community or content from another subject | | | | |
| collect and organize categorical or discrete primary data and display the data in charts, tables and graphs | | | | |
| interpret and draw conclusions from data presented in charts, tables, and graphs | | | | |

**Level 1** - Student rarely applies skills with several errors or omissions.

**Level 2** - Student sometimes applies skills with some errors or omissions.

**Level 3** - Student usually applies skills with few errors or omissions.

**Level 4** - Student consistently applies skills with almost no errors or omissions.

# DAILY MATH WORK RUBRIC

Student ______________________________

| | Level 1 | Level 2 | Level 3 | Level 4 |
|---|---|---|---|---|
| Understanding of Math Concepts | Student demonstrates a limited understanding of math concepts in daily work. | Student demonstrates asatisfactory understanding of skills in daily work. | Student demonstrates a complete understanding of math concepts in daily work.details | Student demonstrates a thorough understanding of math concepts in daily work. |
| Application of Skills Taught | Student rarely applies skills taught in daily work without teacher assistance. | Student applies skills taught in daily work with several errors and omissions. | Student applies skills taught in daily work with few errors and omissions. | Student consistently applies skills taught in daily work with almost no errors and omissions. |
| Math Terminology | Student rarely uses appropriate math terms during math discussions and activities. | Student sometimes uses appropriate math terms during math discussions and activities. | Student usually uses appropriate math terms during math discussions and activities. | Student consistently uses appropriate math terms during math discussions and activities. |
| Preparedness For Class | Student rarely comes prepared with materials and assignments done. | Student sometimes comes prepared with materials and assignments done. | Student usually comes prepared with materials and assignments done. | Student consistently comes prepared with materials and assignments done. |
| Attendance To Task | Student needs frequent reminders to use time wisely. | Student needs some reminders to use time wisely. | Student needs few reminders to use time wisely. | Student rarely needs reminders to use time wisely. |

Additional observations ______________________________

______________________________

______________________________

# CLASS LIST: DAILY MATH WORK

Math Focus:________________________________________

| Studnet Name | Understanding of Concetps | Application of Skills Taught | Math Terminology | Attendance to Task |
|---|---|---|---|---|
| | | | | |
| | | | | |
| | | | | |
| | | | | |
| | | | | |
| | | | | |
| | | | | |
| | | | | |
| | | | | |
| | | | | |
| | | | | |
| | | | | |
| | | | | |
| | | | | |
| | | | | |
| | | | | |
| | | | | |
| | | | | |
| | | | | |
| | | | | |
| | | | | |
| | | | | |
| | | | | |

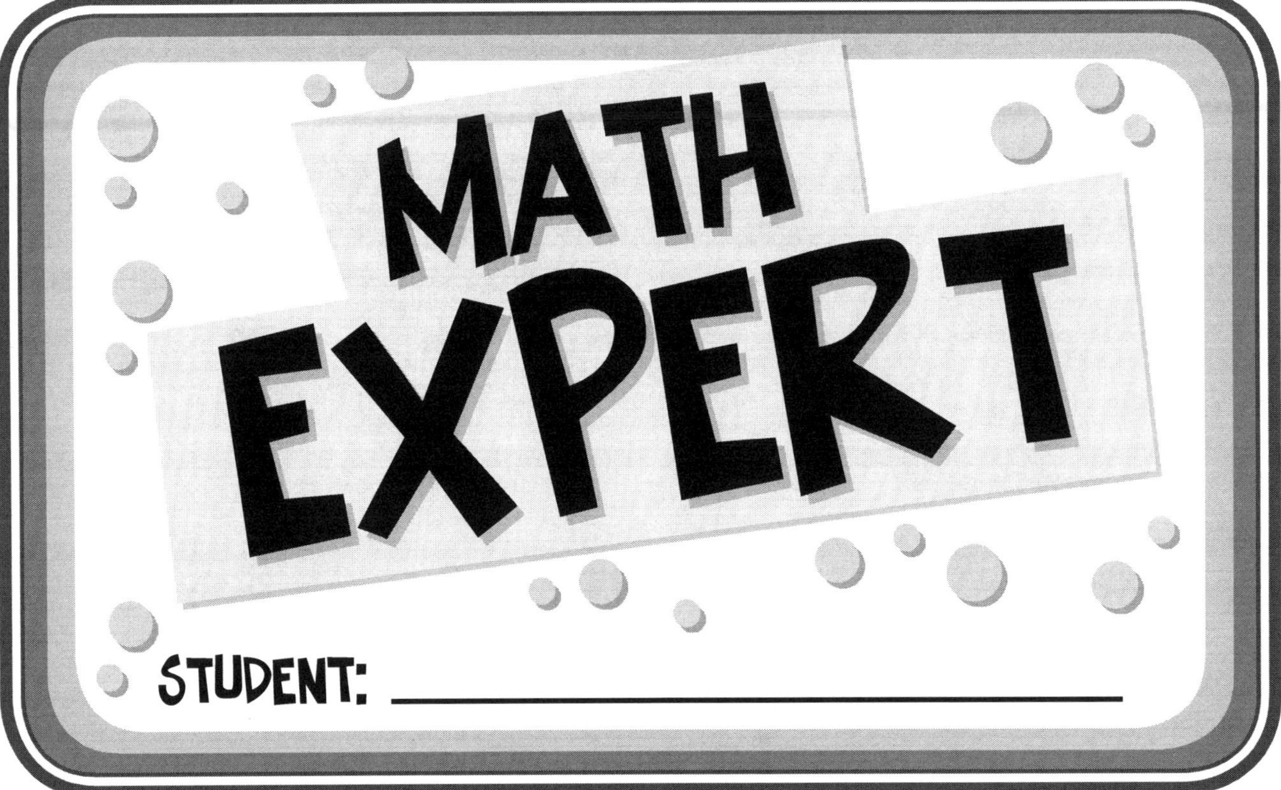


DATA MANAGEMENT

EXPERT!

STUDENT: ______________________

# ANSWER KEY

<table>
<tr><th>PAGE #</th><th>ANSWER KEY</th></tr>
<tr><td>5</td><td> - 6 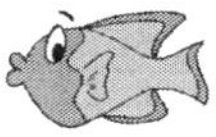 - 10  - 4 1. 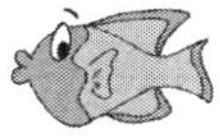 2.  Brain Stretch: a</td></tr>
<tr><td>6</td><td> - 11  - 7  - 5 1.  2.  Brain Stretch: c</td></tr>
<tr><td>7</td><td> 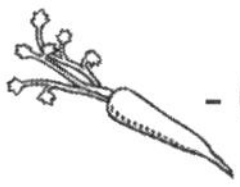- 6  - 2 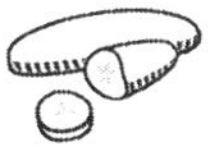 - 9 1. 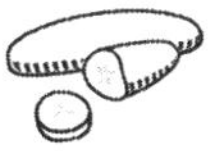 2.  Brain Stretch: b</td></tr>
<tr><td>8</td><td>red-7 blue-13 green-4 purple-8 1. blue 2. green 3. purple<br>4. 20 5. 32 people 6.blue, purple, red, green</td></tr>
<tr><td>9</td><td>cereal - ||| eggs - 𝍸 𝍸 𝍸 pancakes - 𝍸 𝍸 𝍸 𝍸 |<br>grilled cheese - 𝍸 𝍸 𝍸 1. pancakes 2. 24 people<br>3. eggs and grilled cheese 4. 12 people 5.cereal 6. 18 people</td></tr>
<tr><td>10</td><td>swimming-10 baseball -9 soccer-11 going to the beach-7</td></tr>
<tr><td>11</td><td>pennies-15 nickels-9 dimes-16 quarters-6 loonies-5</td></tr>
<tr><td>12</td><td>fish-3 bird-1 cat-2 dog-3 1. dog and fish 2. bird</td></tr>
<tr><td>13</td><td>juice -3 lemonade -2 soda -1 milk - 4 Brain Stretch: 1. juice 2. none</td></tr>
<tr><td>14</td><td>1. Friday 2. 12 fish 3. Monday 4. 15-3=12 5. 15 fish 6. 18</td></tr>
<tr><td>15</td><td>1. Mr. Poulos 2. 22 3. Ms. Chin, Mr. Patel 4. Ms. Apor<br>5. Ms. Apor, Ms. Chin 6. Mr. Poulos</td></tr>
<tr><td>16</td><td>1.  - 2  - 4  - 5 2.  3. </td></tr>
<tr><td>17</td><td>1.  - 4  - 6  - 3 2.  3.  Brain Stretch: 11, 8</td></tr>
</table>

# ANSWER KEY

| PAGE # | ANSWER KEY |
|---|---|
| 18 | 1. 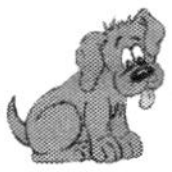 - 6 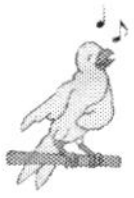 - 2 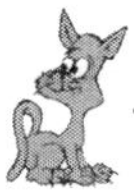 - 5 2. 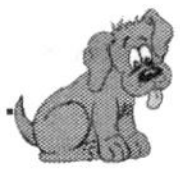 3. 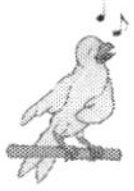 Brain Stretch: 10, 4 |
| 19 | 1. 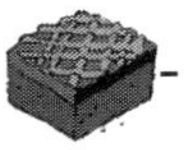 - 4 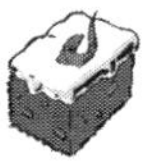 - 3 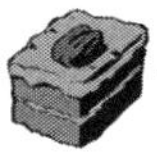 - 6 2. 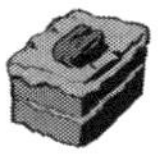 3. 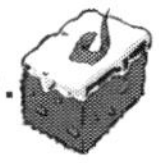 4.13 people |
| 20 | 1. 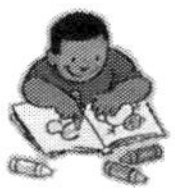 - 3 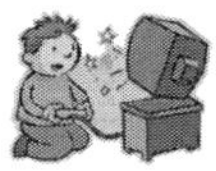 - 4 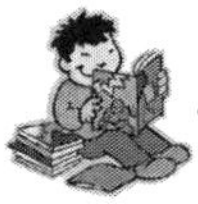 - 2 2. 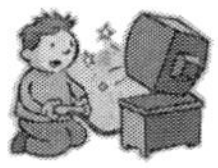 3. 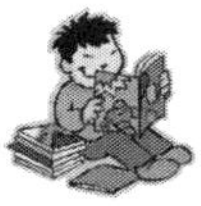 4. 9 people |
| 21 | 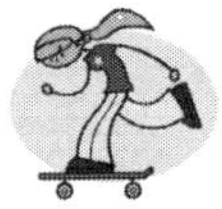 𝍸 \|\|\| 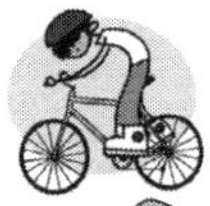 𝍸 𝍸 𝍸 \| 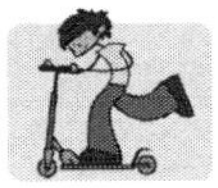 𝍸 𝍸 𝍸 𝍸<br>1. 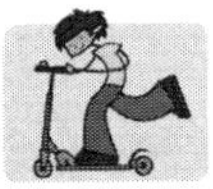 2. 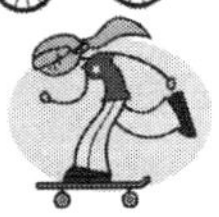 3. 8 people 4. 4 people |
| 22 | 1. 8 2. 6 3. 16 4. 4 5. 8 6. 4 |
| 23 | 1. lemonade 2. milk 3. 30 4. 50 5. 70 6.150 7. 40 8. 80 |
| 24 | 1. strawberry 2. chocolate 3.40 4. 20 5. 84 6. 20 7.24 8. 20 9. 44 10. 128 |
| 25 | 1. 12 2. 6 3. 3 4. sugar 5. 3 6. sugar, chocolate chip, oatmeal raisin, gingerbread |
| 26 | 1. Tom 2. 10 3. 40 4. 15 5. 25 6. 140 |
| 27 | 1. circle 2. star Brain Stretch: 12, 4 |
| 28 | 1. 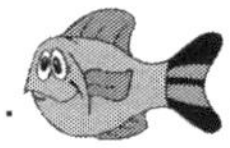 - 3 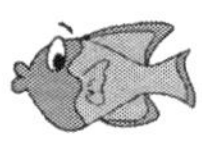 - 7 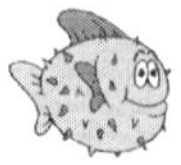 - 5 2. 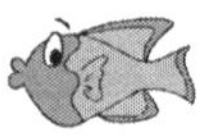 3. 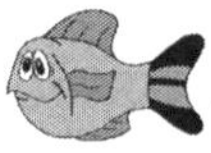 4.15 |
| 29 | 1. 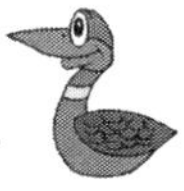 - 7 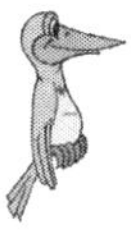 - 3 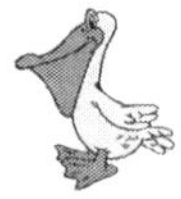 - 8 2. 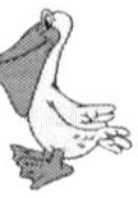 3. 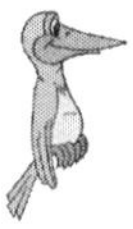 4.18 |

# ANSWER KEY

| PAGE # | ANSWER KEY |
|---|---|
| 30 | 1.  - 2  - 4  - 3 2.  3.  Brain Stretch: 9, ,7 |
| 31 | 1.  - 5  - 7  - 3 2.  3.  Brain Stretch:13, 3 |
| 32 | 1.drums-7 guitar-4 harmonica-6 2. drum 3. guitar 4. 11 5.2 |
| 33 | 1. a. mall b. park c. circus d. zoo e. library 2. 6 3. 8 4. mall 5. library |
| 34 | 1. sandwich 2. noodles 3. 9 4. 8 5. 3 6. soup |
| 35 | 1. money 2. toys 3. 12 4. 12 5. 4 6. clothes |
| 36 | 1.Shelley 2.$40 3.Vicki 4.$10 5.$150 6.Rob |
| 37 | 1.peach 2.apple 3.pear, cherry 4.3 5.7 6.29 |
| 38 | 1.a. bananas b. grapes c. apples 2. 2 3.10 4.apples 5.22 |
| 39 | 1.a. hockey card b. doll c. stuffed toy d. stamp e.shell 2. hockey card 3.shell 4.16 5.10 |
| 40 | 1.Thursday 2. 35 3.Wednesday 4.Tuesday 5.10 6. Monday |
| 41 | 1. Nicolette 2. Sebastian 3. 30 4. 130 apples 5. 10 6. 260 |
| 42 | 1. tulips 2. roses 3. 4 4. 4 5. 8 6. 64 |
| 43 | 1. answers will vary 2. Tara, Sophie 3. 2 4. 3 5. 40 6. Rich |
| 44 | 1. 18 2. zebra 3. lion 4. zebra, flamingo, giraffe, lion |

# ANSWER KEY

| PAGE # | ANSWER KEY |
|---|---|
| 45 | 1.chips 2. 6 3. 14 children 4. chips 5. pretzels<br>6. carrots, chips, cookies, pretzels |
| 46 | 1. potatoes, broccoli, carrots, peas 2. 2 3. 4 4.potatoes 5. peas |
| 47 | 1. 9 2. silver 3. 4 4. 20 5. 5 6. 4 |
| 48 | 1. summer, spring, winter, autumn 2. 1 3. spring 4. 25 5. summer |
| 49 | see graph Brain Stretch: answers will vary |
| 50 | 1. winter, summer, spring, fall 2. 10 3. 25 4. autumn 5. winter 6. 5 |
| 51 | Answers will vary. |
| 52 | Answers will vary. |
| 53 | 1. Monday 2. February 5th 3. 4 4. February 14th 5.Thursday<br>6. Friday 7. February 1 8. 4 9. February 18th 10. February 6th |
| 54 | 1. Tuesday 2. 5 3. March 12th 4. March 2nd 5. Saturday 6. March 17th<br>7. Monday 8. 4 9. March 4th 10. 4 |
| 55 | 1. Monday 2. Monday 3. August 19th 4. August 4th 5. 5 6.Saturday<br>7. August 7th 8. August 30th 9. 5 10. 4 |
| 56 | 1. January 5th 2. Wednesday 3. 5 4. January 18th 5. Friday<br>6. 5 7. January 7th 8. Thursday 9.January 24th 10. Monday |
| 57 | 1. recycling, dance, swimming, crafts, school newspaper, chess<br>2. band, track and field, crafts, school newspaper, chess 3. crafts, school newspaper, chess Brain Stretch: 30 |
| 58 | 1. Nicole, Ross, David 2. Scott, Sophie, Chris 3. Paul, Elizabeth, Mary<br>Brain Stretch:12 |
| 59 | 1. Ben, Megan, Lisa 2. Spencer, Madelyn, David 3. Kaitlyn, Michael<br>Brain Stretch: 16 |
| 60 | Answers will vary. |